# MOUSE LIVER
NEOPLASIA

## CHEMICAL INDUSTRY INSTITUTE OF TOXICOLOGY SERIES

**Gibson**: Formaldehyde Toxicity

**Golberg**: Structure-Activity Correlation as a Predictive Tool in Toxicology: Fundamentals, Methods, and Applications

**Gralla**: Scientific Considerations in Monitoring and Evaluating Toxicological Research

**Popp**: Mouse Liver Neoplasia: Current Perspectives

### Forthcoming

**Barrow**: Toxicology of the Nasal Passages

**Hamm**: Complications of Viral and Mycoplasmal Infections in Rodents to Toxicology Research and Testing

**Rickert**: Toxicity of Nitroaromatic Compounds

# MOUSE LIVER NEOPLASIA
Current Perspectives

*Edited by*

## James A. Popp
*Chemical Industry Institute of Toxicology*

## ●HEMISPHERE PUBLISHING CORPORATION
Washington    New York    London

DISTRIBUTION OUTSIDE THE UNITED STATES
## McGRAW-HILL INTERNATIONAL BOOK COMPANY

Auckland   Bogotá   Guatemala   Hamburg   Johannesburg   Lisbon
London   Madrid   Mexico   Montreal   New Delhi   Panama   Paris
San Juan   São Paulo   Singapore   Sydney   Tokyo   Toronto

**MOUSE LIVER NEOPLASIA: Current Perspectives**

Copyright © 1984 by Hemisphere Publishing Corporation. All rights reserved. Printed in the United States of America. Except as permitted under the United States Copyright Act of 1976, no part of this publication may be reproduced or distributed in any form or by any means, or stored in a data base or retrieval system, without the prior written permission of the publisher.

1 2 3 4 5 6 7 8 9 0   B R B R   8 9 8 7 6 5 4

Library of Congress Cataloging in Publication Data
Mouse liver neoplasia.

  (Chemical Industry Institute of Toxicology series)
  Includes bibliographical references and index.
  1. Liver–Cancer–Animal models. 2. Mice–Diseases.
3. Mice–Physiology. I. Popp, James A., date.
II. Series. [DNLM: 1. Liver neoplasms. 2. Neoplasms,
Experimental. WI 735 C976]
RC280.L5C87   1984      616.99'36027      84-3767
ISBN 0-89116-300-X
ISSN 0278-6265

# Contents

| | |
|---|---|
| Contributors | ix |
| Preface | xi |

**CHAPTER 1**   Morphology of Potential Preneoplastic Hepatocyte Lesions and Liver Tumors in Mice and a Comparison with Other Species    1
*J. M. Ward*

| | |
|---|---|
| Introduction | 1 |
| Classification and Histogenesis | 1 |
| Hepatocellular Adenoma | 6 |
| Hepatocellular Carcinoma Arising in Hepatocellular Adenoma | 8 |
| Hepatocellular Carcinoma | 9 |
| Induction of Unique Tumor Types in Mice by Chemical Carcinogens | 11 |

|   |   |   |
|---|---|---|
|  | Other Hepatic Tumors in Mice | 15 |
|  | Summary of Comparative Aspects | 18 |
|  | References | 19 |
| CHAPTER 2 | Occurrence of Mouse Liver Neoplasms in the National Cancer Institute Bioassays<br>*Thomas E. Hamm, Jr.* | 27 |
|  | References | 37 |
| CHAPTER 3 | Implications of Mouse Liver Neoplasms for Predicting Carcinogenicity<br>*Robert A. Squire* | 39 |
|  | References | 44 |
| CHAPTER 4 | Genetic, Hormonal, and Dietary Factors in Determining the Incidence of Hepatic Neoplasia in the Mouse<br>*Paul Grasso* | 47 |
|  | Introduction | 47 |
|  | Effects of Diet | 48 |
|  | Genetic Factors | 51 |
|  | Hormonal Factors | 53 |
|  | Discussion and Summary | 55 |
|  | References | 58 |
| CHAPTER 5 | Kinetics of Induction and Growth of Basophilic Foci and Development of Hepatocellular Carcinoma by Diethylnitrosamine in the Infant Mouse<br>*Stan D. Vesselinovitch and N. Mihailovich* | 61 |
|  | Introduction | 61 |
|  | Materials and Methods | 65 |
|  | Results | 65 |
|  | Discussion | 75 |
|  | Summary and Conclusions | 79 |
|  | References | 80 |
| CHAPTER 6 | Comparison of Morphologic and Biologic Characteristics of Liver Tumors in Control Mice and Mice Treated with 2-Acetylaminofluorene or Benzidine Dihydrochloride<br>*Charles H. Frith* | 85 |
|  | Introduction | 85 |
|  | Materials and Methods | 85 |

|  |  |  |
|---|---|---|
|  | Results | 86 |
|  | Discussion | 91 |
|  | References | 93 |
| CHAPTER 7 | DNA Alkylation and Cell Replication and Mouse Liver Carcinogenesis<br>*James A. Swenberg and Charles Lindamood III* | 95 |
|  | Introduction | 95 |
|  | DNA Alkylation and Repair | 95 |
|  | Cell Replication in Initiation and Promotion | 96 |
|  | Cell Specificity in Hepatocarcinogenesis | 97 |
|  | References | 102 |
| CHAPTER 8 | Potential Use of Initiation-Promotion Studies in Understanding Mouse Liver Neoplasia<br>*James A. Popp and Thomas B. Leonard* | 107 |
|  | References | 113 |
| CHAPTER 9 | Mouse Hepatic Neoplasia: Differences among Strains and Carcinogens<br>*Boris H. Ruebner, M. Eric Gershwin, S. W. French, Earl Meierhenry, Pat Dunn, and Lucy S. Hsieh* | 115 |
|  | Introduction | 115 |
|  | Materials and Methods | 115 |
|  | Results | 116 |
|  | Discussion | 137 |
|  | References | 141 |
| CHAPTER 10 | Characterization of Spontaneous and Chemically Induced Mouse Liver Tumors<br>*Frederick F. Becker* | 145 |
|  | Introduction | 145 |
|  | Materials and Methods | 146 |
|  | Transplantation | 147 |
|  | Results | 147 |
|  | Virus Studies | 151 |
|  | Discussion | 154 |
|  | References | 156 |

| CHAPTER 11 | Reversibility, Persistence, and Progression of Safrole-induced Mouse Liver Lesions following Cessation of Exposure | 161 |
|---|---|---|
| | *Michael M. Lipsky, David C. Tanner, David E. Hinton, and Benjamin F. Trump* | |
| | Introduction | 161 |
| | Materials and Methods | 162 |
| | Results | 163 |
| | Discussion | 171 |
| | References | 175 |
| Index | | 179 |

# Contributors

FREDERICK F. BECKER
The University of Texas System
 Center
Houston, Texas

PAT DUNN
University of California
Davis, California

S. W. FRENCH
University of California
Davis, California

CHARLES H. FRITH
Intox Laboratories
Little Rock, Arkansas

M. ERIC GERSHWIN
University of California
Davis, California

PAUL GRASSO
Group Occupational Health Center
British Petroleum
Sunbury on Thames, Middlesex,
 England

THOMAS E. HAMM, Jr.
Chemical Industry Institute
 of Toxicology
Research Triangle Park, North Carolina

DAVID E. HINTON
University of West Virginia Medical
 Center
Morgantown, West Virginia

LUCY S. HSIEH
University of California
Davis, California

THOMAS B. LEONARD
Chemical Industry Institute
 of Toxicology
Research Triangle Park, North Carolina

CHARLES LINDAMOOD III
Chemical Industry Institute
 of Toxicology
Research Triangle Park, North Carolina

MICHAEL M. LIPSKY
University of Maryland Medical School
Baltimore, Maryland

EARL MEIERHENRY
University of California
Davis, California

N. MIHAILOVICH
University of Chicago
Chicago, Illinois

JAMES A. POPP
Chemical Industry Institute
 of Toxicology
Research Triangle Park, North Carolina

BORIS H. RUEBNER
University of California
Davis, California

ROBERT A. SQUIRE
Johns Hopkins University
Baltimore, Maryland

JAMES A. SWENBERG
Chemical Industry Institute
 of Toxicology
Research Triangle Park, North Carolina

DAVID C. TANNER
University of Maryland Medical School
Baltimore, Maryland

BENJAMIN F. TRUMP
University of Maryland Medical School
Baltimore, Maryland

STAN D. VESSELINOVITCH
University of Chicago
Chicago, Illinois

J. M. WARD
National Cancer Institute
Frederick, Maryland

# Preface

This book reviews the available published and unpublished experimental data on the biology of mouse liver neoplasms. Although this tumor is frequently found in bioassay testing of drugs and chemicals, its biology and development is poorly understood. This critical lack of information must be overcome before the meaning of a positive tumor response in the mouse liver can be adequately evaluated for human risk assessment. By summarizing the current state of knowledge concerning mouse liver neoplasms, this book identifies many of the deficiencies in the current knowledge base and thereby identifies the critical areas for future research.

*James A. Popp*

Chapter 1

# Morphology of Potential Preneoplastic Hepatocyte Lesions and Liver Tumors in Mice and a Comparison with Other Species

**J. M. Ward**

INTRODUCTION

Over the past 10 years, an increasing number of chemicals have been found to be associated with an elevation of liver tumor incidence in mice after routine 2-year or lifetime carcinogenesis bioassays (1-45). These liver carcinogens represent many chemical classes, including aromatic amines, halogenated compounds, cyclic ethers, inorganics, thioureas, polycyclic aromatic hydrocarbons, nitrosamides, nitrosamines, azoxymethanes, phthalates, hormones, amides, lactones, and methylenedioxy compounds (43). In addition to the chemically-induced tumors, naturally occurring liver tumors are found in high incidence in some strains of mice (46-57) and can be modulated by various factors (7) including hormones, diet, and genetics (49). As a result, the mouse liver tumor has been criticized as an inappropriate endpoint in carcinogenesis bioassays (6,9). This chapter reivews the pathology of hepatocellular and other liver tumors in mice and compares them to similar lesions in other species.

CLASSIFICATION AND HISTOGENESIS

A classification of mouse hepatic lesions has been proposed (Table 1) and is in current use by the National Cancer Institute, National Center for Toxicological Research and the National Toxicology Program (12). The classification is based on histogenetic and morphologic features similar to those found in rats (60-61) and other species. The histogenesis of naturally occurring hepatocellular tumors in mice has been suggested by several authors (2,3,12,55,59,60). Naturally occurring hepatocellular tumors and

---

The author is grateful for the editorial assistance of Kevin Beall and photography of Larry Ostby and Gary Best, and to medical illustrator Joan McClellan.

TABLE 1. A Classification of Potential Preneoplastic and Neoplastic Liver Lesions in Mice[a]

---

Potential preneoplastic hepatocellular lesions

1. Focus of altered cells (focal hyperplasia, focal cellular alteration)

    A. Basophilic
    B. Eosinophilic
    C. Vacuolated
    D. Clear
    E. Mixed

Neoplasms

1. Hepatocellular

    A. Hepatocellular adenoma
        Basophilic, eosinophilic, vacuolated, clear, mixed

    B. Focal hepatocellular carcinoma in adenoma

    C. Hepatocellular carcinoma

        By pattern:                By differentiation:

        Trabecular                 Well differentiated
        Solid (nontrabecular)      Moderately well differentiated
        Adenocarcinoma             Poorly differentiated

    D. Hepatoblastoma

2. Biliary

    Cholangioma, cholangiocarcinoma
    Mixed cholangiocarcinoma-hepatocellular carcinoma

3. Vascular

    Hemangioma, hemangiosarcoma

4. Histiocytic

    Histiocytic sarcoma (Kupffer cell sarcoma)

5. Metastatic tumors

---

[a]Modified from Frith and Ward. (12).

Some chemically induced tumors in mice follow a sequence of development similar to that in rats (Fig. 1) (58,61). While many questions and controversies remain concerning the histogenesis of mouse tumors (7), similar controversies have also arisen for liver lesions in rats (61), other animals (62), and humans (63-67).

The naturally occurring mouse tumors usually arise as foci of basophilic hepatocytes (12,55,60), which can be either larger or smaller than normal hepatocytes. They may also be composed of eosinophilic (Fig. 2), clear, or vacuolated hepatocytes (Fig. 3). These foci have also been reported as focal hyperplasia and focal cellular alteration. Induced foci may be morphologically similar or different compared to foci in control animals. In some experiments, basophilic hepatocytes with increased glycogen (clear portions of the cytoplasm) appear to represent the earliest stage of neoplastic development (2,37). Areas of cellular alteration as noted in rats may also be seen in mice. The type 1 nodule (7) and hyperplastic nodule (5,7,12,41), described by some authors, appear to represent "areas" in mice. Little compression of adjacent normal parenchyma is seen, and portal areas may remain. There is evidence that some foci, induced by safrole in the diet, may regress, but most natural and induced foci progress and become larger. Foci may be identified and characterized by a variety of special staining procedures.

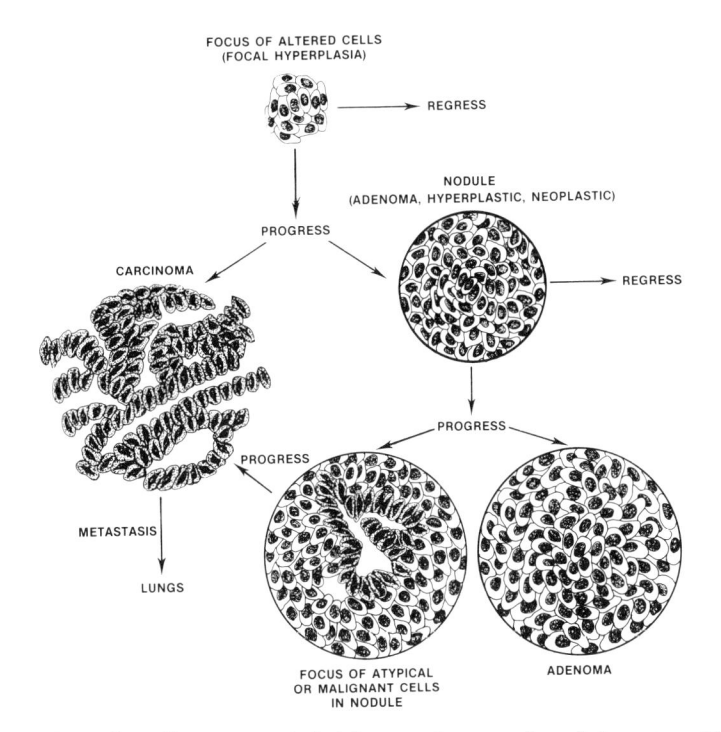

FIGURE 1. The potential histopathogenesis of hepatocellular tumors in mice and other species.

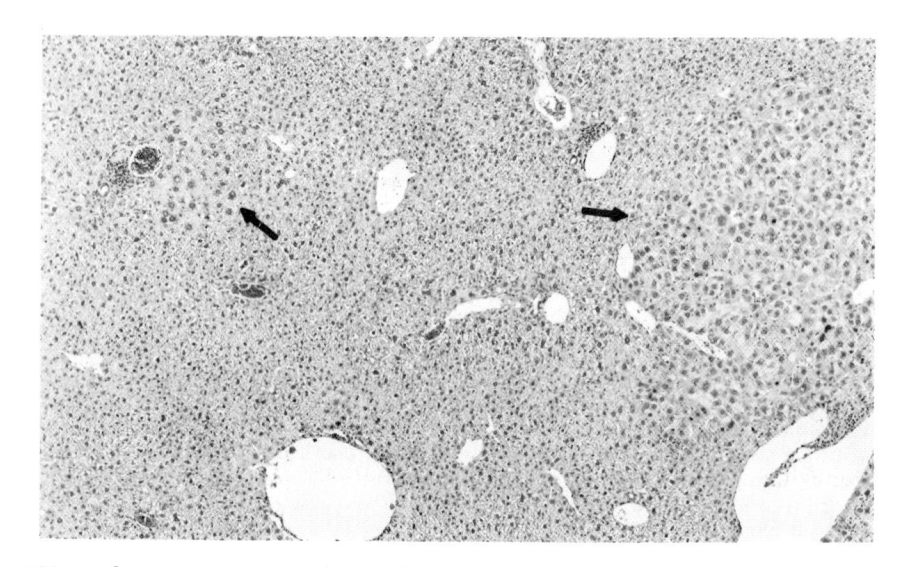

FIGURE 2. Two hepatocellular foci composed of eosinophilic hepatocytes (arrows) larger than normal hepatocytes in liver of mouse exposed to nitrofen, X54. (Courtesy of Dr. Karen Hoover.)

The hepatocytes in foci or tumors may be iron-negative after induced siderosis (23,56). A recent study demonstrated that hepatocytes in foci, induced by a single intraperitoneal injection of diethylnitrosamine in neonatal mice, did not contain glucose-6-phosphatase activity, and the foci were dose-sex- and time-related (27). Naturally occurring foci and tumors do not stain for gamma-glutamyl transpeptidase activity (19,57). Other

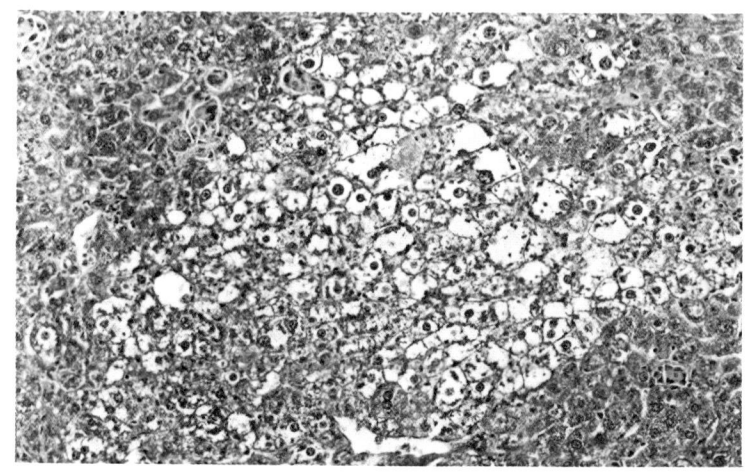

Figure 3. Focus of vacuolated and clear hepatocytes in a mouse, X130.

normal hepatic enzyme activity may be altered in cells of the foci (9,14,22,36). The comparative aspects of foci and liver tumors in four species are shown in Table 2. Dysplastic foci, seen in monkeys (Fig. 4) and humans (63,65,67), are infrequent in mice.

Hepatocellular foci in mice may represent potential preneoplastic lesions (12,55,60). The morphology of hepatocytes in foci resembles that in hepatocellular adenomas, the next stage in potential neoplastic progression. Histochemically, foci may be similar to adenomas (19,21,36). For example, after induced siderosis, foci and adenomas may be iron-deficient (21,56). Limited evidence has been presented that foci in livers of mice exposed to some carcinogens may transform directly to carcinoma without adenoma formation (Fig. 1) (14,33).

TABLE 2. Comparative Etiology and Pathology of Natural Hepatocellular Tumors

| Characteristic | Mouse | Rat | Monkey | Human | Selected references |
|---|---|---|---|---|---|
| Etiology | | | | | |
| Genetics | + | − | − | − | 49 |
| Virus | − | − | − | + | 65 |
| Chemicals | − | − | − | + | 64,69 |
| Pathology | | | | | |
| Cirrhosis in livers with tumors | − | − | − | + | 64,65 |
| Major potential preneoplastic focus | | | | | |
| Basophilic | + | + | | | 11,12,61 |
| Clear | | | + | | 38 |
| Dysplastic | | | | + | 63,65,67 |
| Enzyme alterations | + | + | + | + | 14,22,61,67 |
| Benign nodules | + | + | + | + | 38,41,60,61,66 |
| Carcinoma in nodules | + | + | − | − | 23,60,61 |
| Carcinoma | + | + | + | + | 12,41,61,64 |

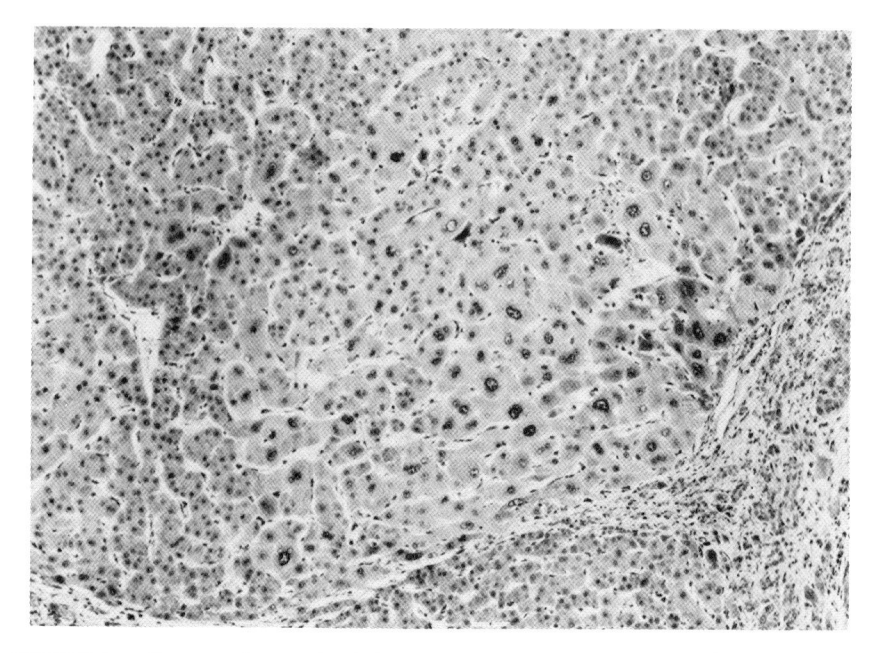

FIGURE 4. Focus of atypical hepatocytes in liver of monkey exposed to aflatoxin $B_1$. Similar foci are rarely seen in mice, but are seen more commonly in humans, X80.

HEPATOCELLULAR ADENOMA

When foci become the size of a liver lobule, they may be seen grossly as pearly nodules (Fig. 5). Compression of adjacent parenchyma is usually observed either on one edge or around the entire lesion (Figs. 6-9). Liver plates are 1-2 cells thick and regular, although the lesion appears nontrabecular or solid (Fig. 7). Tumor edges may be irregular and show so-called microinvasion (Fig. 10) (4,47). At this state of development, the lesion has been described by more than 10 different diagnoses, as listed in Table 3. The terminology used for this lesion depends, in part, on one's experience, bias, degree of diagnostic conservatism, and lack of information on the biology of the lesion. Hepatocellular adenoma appears to be an appropriate designation for the lesion, because the lesion is morphologically well differentiated and has general features associated with benign tumors in other organs of mice and other species (see Table 4). In mice, the benign appearance of this lesion at a given time does not suggest that it will stay benign for the life of the mouse. Instead, at least in control mice, some adenomas enlarge and develop foci of atypical or malignant cells (similar to those in trabecular carcinoma) within them (Figs. 11,12) (55,60). In control B6C3F1 mice, up to 50% of the small adenomas may progress to carcinomas (60), indicating that the adenoma in mouse liver may be a stage in the development of carcinoma (59). Similar progression has been

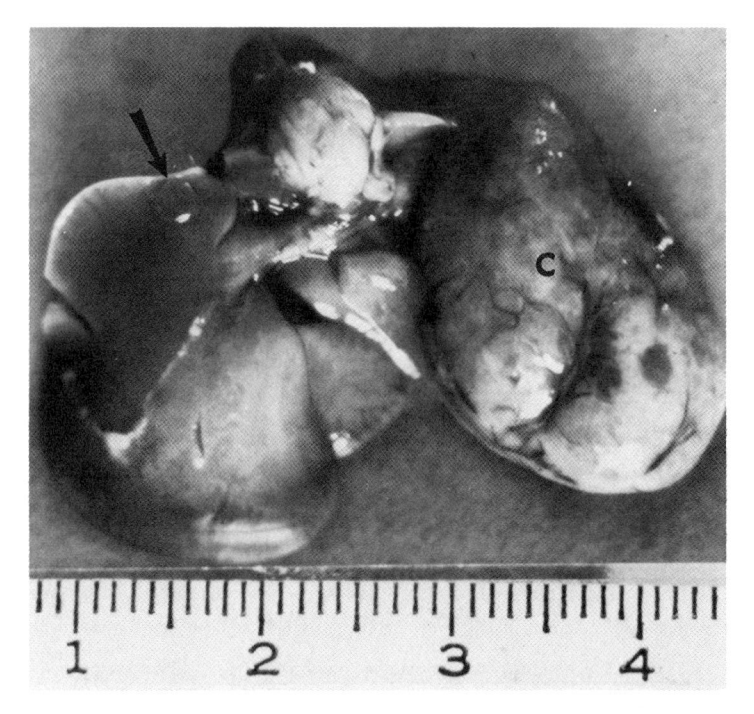

FIGURE 5. Gross illustration of small (early) hepatocellular adenoma (arrow) and large hepatocellular carcinoma (C) in a control B6C3F1 mouse at 2 years of age.

observed in the development of hepatic neoplasms in rats (Table 5) (61).

The argument has been made that mouse adenomas are not neoplastic, but rather hyperplastic. Limited evidence has been presented that under certain conditions, e.g., exposure to dieldrin (36) or γ-benzene hexachloride (16), some liver nodules in mice appear to regress. Arguments against this point of view include items in Table 4 and the fact that partial hepatectomy and chronic hepatic damage in mice lead to a diffuse increase in the number of mitotic figures in hepatocytes throughout the liver (J. Ward, unpublished observations) (Fig. 13), rather than a localized increase in mitotic figures, indicating the development of regenerative nodules. Large compressing nodules develop in chronically damaged mouse liver, but only months after the increase in mitotic figures are seen. These findings suggest that regeneration in mouse liver is similar to that in other species and occurs by a diffuse increase in cell proliferation rather than by nodular hyperplasia. However, regenerative nodular lesions may be seen in cirrhotic mouse livers and may be very difficult to distinguish from adenomas. Further characterization of lesions in cirrhotic livers is needed. Thus, any nodular lesion, at least in noncirrhotic livers, would appear to be a neoplasm. Transplantation has been accomplished for a low

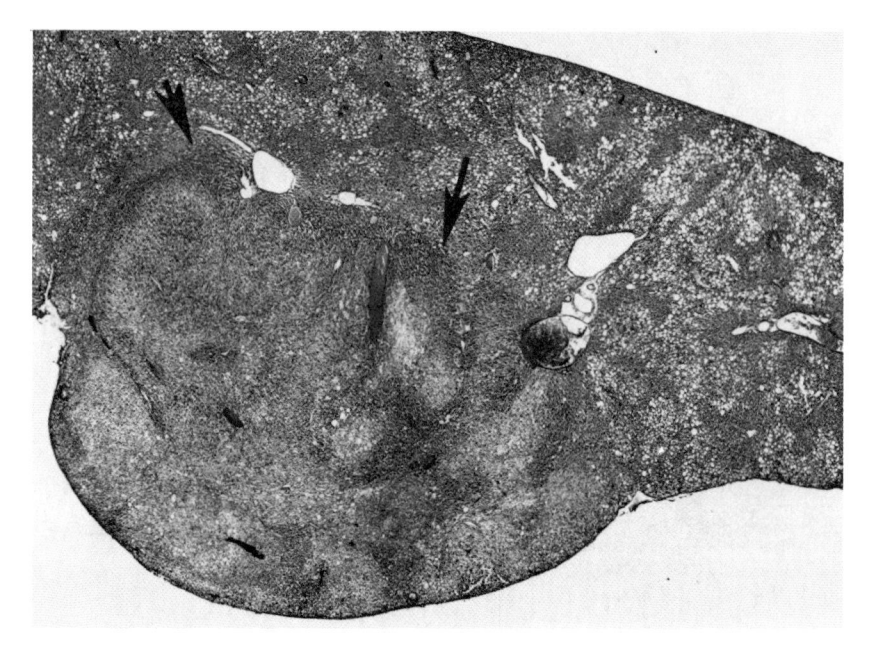

FIGURE 6. Hepatocellular adenoma with basophilic hepatocytes on edge of tumor (arrows) and vacuolated hepatocytes toward center of tumor, X16.

percentage of adenomas (7,56). This finding is evidence for autonomous progressive growth (neoplasia), at least for the transplantable tumors, but not malignancy, since some benign rat tumors are transplantable and maintain their benign morphology after transplantation (69).

HEPATOCELLULAR CARCINOMA ARISING IN HEPATOCELLULAR ADENOMA

In B6C3F1, C3H, and other strains of mice, perhaps 40% of the large adenomas have foci of prominent trabecular formations (nodule in nodule, focal atypia, focal carcinoma) within lhe adenoma (Figs. 11,12) (46,55,60). The trabeculae are usually several cells thick. Ruebner et al. have shown, at least for some trabecular foci, that the glucose-6-phosphate dehydrogenase enzyme pattern differs from that in adenomas (36). Atypical foci in adenomas have also been seen in mice exposed to either benzidine (11) or safrole (23). The trabecular foci or areas of atypia are almost never seen in situ or in small nodules in control mouse livers. It appears that this second population of more malignant cells arises within the adenoma, proliferates faster than adenoma cells, and eventually outgrows the adenoma. Some adenomas grow quite large (greater than 1 cm in diameter) and do not seem to develop trabecular foci, even if the mouse is allowed to live its

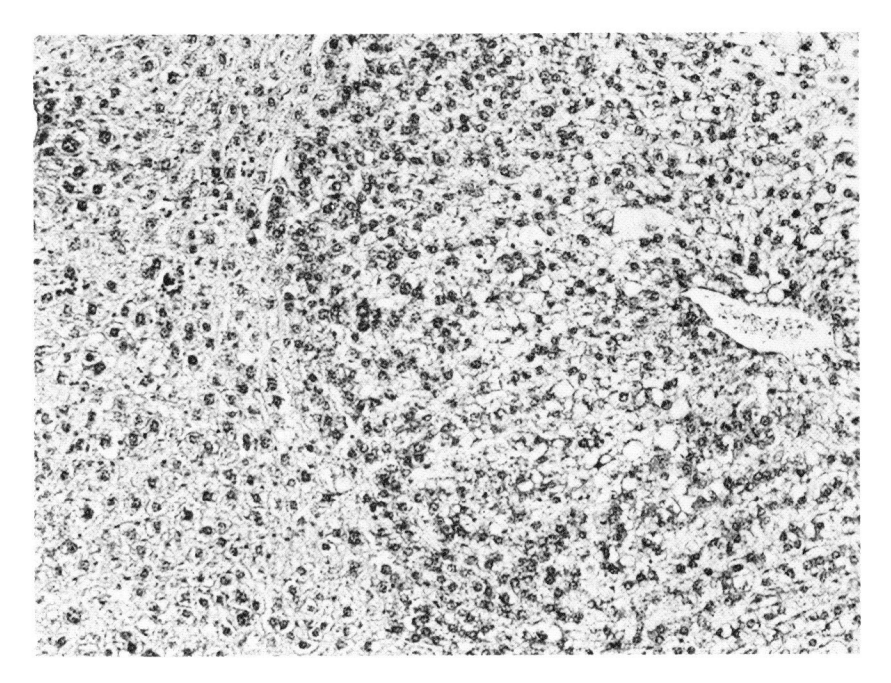

FIGURE 7. Portion of a hepatocellular adenoma in B6C3F1 mouse. Note compression of adjacent normal hepatocytes by the expansive adenoma and basophilic and clear cell appearance of the hepatocytes in the adenoma, X130.

natural lifespan. These lesions would appear to be classical adenomas and not stages in the development of carcinoma.

HEPATOCELLULAR CARCINOMA

Grossly, carcinomas are large and have prominent blood vessels. In control mice with a low incidence of liver tumors they are usually single (Fig. 5), but are frequently multiple in mice exposed to a carcinogen (Fig. 14). Histologically, hepatocellular carcinomas may be classified by pattern and degree of differentiation (Table 1). However, the ultimate evidence of malignancy is metastasis. In mice, hepatocellular tumors that metastasize invariably have a prominent trabecular pattern in the primary tumor (Fig. 15) (19,41,60). Frith et al. (13) have shown that the incidence of pulmonary metastases is related to tumor size and degree of differentiation. Most hepatocellular tumors metastasize only when more than 8 mm in diameter, a tumor size in which trabecular foci have developed. Poorly differentiated tumors metastasize more frequently than well differentiated tumors. In comparison with other species, the metastatic rate of liver tumors varies with the carcinogen, species, and tumor type (Tables 5 and 6). Few naturally occurring monkey or rat liver tumors metastasize,

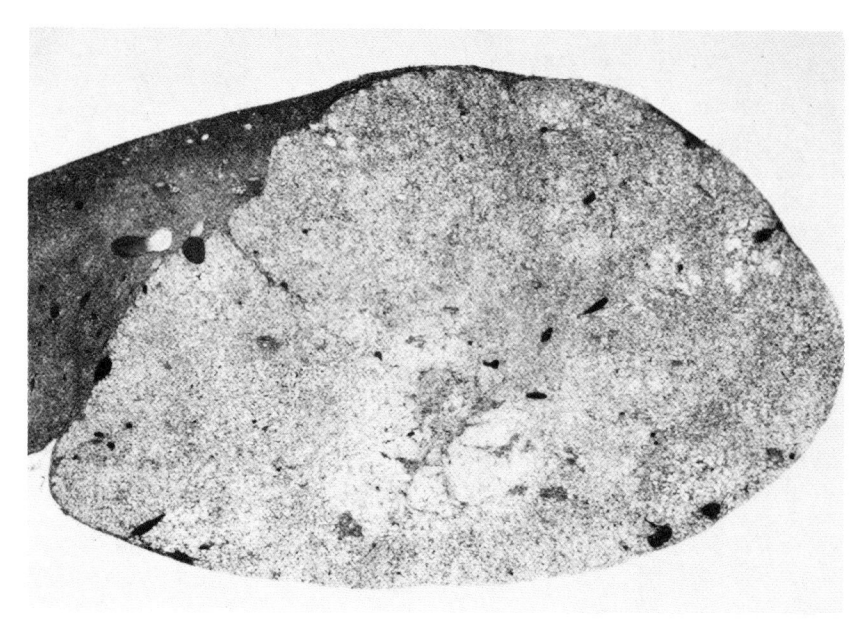

FIGURE 8. Hepatocellular adenoma composed predominantly of vacuolated hepatocytes, X20.

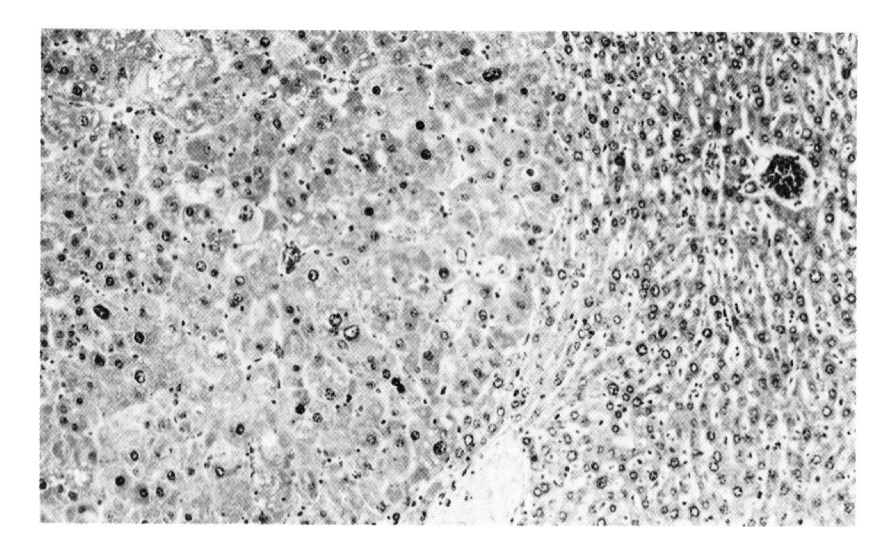

FIGURE 9. Portion of hepatocellular adenoma composed of eosinophilic hepatocytes, X130. (Courtesy of Dr. Charles Frith.)

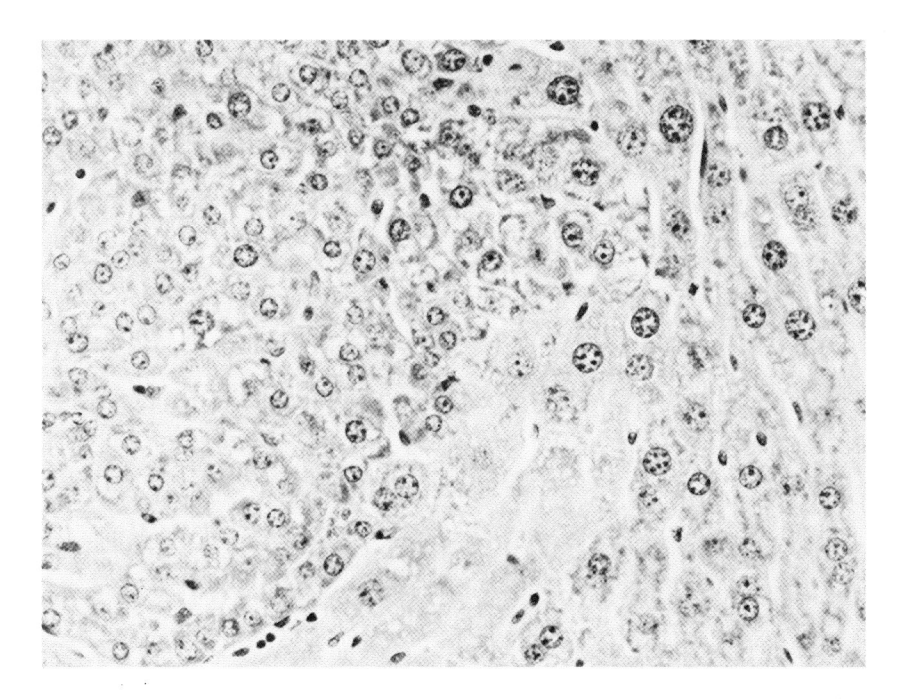

FIGURE 10. Potential microinvasion on the edge of a hepatocellular adenoma in a B6C3F1 mouse, X330.

because most such tumors are benign. Mouse liver tumors appear to metastasize with the same frequency as those in other species.

INDUCTION OF UNIQUE TUMOR TYPES IN MICE BY CHEMICAL CARCINOGENS

Using the classification scheme suggested by Frith and Ward (12), some carcinogens have been shown to induce unique morphologic and biologic types of liver tumors in mice (Figs. 16,17). In untreated B6C3F1 mice, 80% of the adenomas are composed of hepatocytes with basophilic cytoplasm, while only 20% have eosinophilic cytoplasm (60). In these mice, carcinomas are usually trabecular tumors composed of basophilic cells. In contrast, eosinophilic foci and tumors (15,16,36,39) may be the primary type of lesion induced by certain carcinogens (Table 6). The eosinophilia of the cytoplasm may be due to the presence of smooth endoplasmic reticulum and/or mitochrondria (36). Mallory bodies in tumor cells are induced by dieldrin in mice (26). Unique tumors have been seen in livers and other tissues of rats (68) and other species (Table 7).

Using the classification system of Weisburger and Williams (45), some epigenetic carcinogens cause one of the two types of liver tumors (Table 6). For these chemicals, there is no relationship between type of carcinogen, type of tumor, and metastatic rate. These findings suggest that liver tumor classification may play no

TABLE 3. Terms Used for Morphologically Similar Mouse Liver Lesions

| Term | References |
| --- | --- |
| Infer neoplasia | |
|     Hepatocellular adenoma | 12,25,35,55,60 |
|     Adenomatous nodule | 41 |
|     Hepatoma | 18,46,52 |
|     Benign hepatoma | 12 |
|     Neoplastic nodule | 12 |
|     Liver (hepatic) tumor (neoplasm) | 56 |
|     Hepatocellular carcinoma (low-grade malignancy) | 59 |
| Infer hyperplasia | |
|     Simple nodule | 5,7 |
|     Type A nodule | 7,42 |
|     Type 1 or 2 nodule | 7 |
|     Hyperplastic nodule | 6,7,41,48,51 |
|     Nodular hyperplasia | 7 |
|     Liver (hepatic) nodule | 3,4,5,9,14,21,26,47 |

FIGURE 11. Prominent trabecular foci (arrows) (hepatocellular carcinoma) within a hepatocellular adenoma, X115.

TABLE 4. Evidence That the Mouse Hepatocellular Adenoma Lesion Is Neoplastic

| Characteristics of adenoma | Neoplastic characteristic | References |
| --- | --- | --- |
| Morphology | Identical to benign tumors in other organs and species | 12,55,60 |
| Progressive growth and irreversibility | Present | 20,23,36,55 |
| Transplantability | Autonomous growth (neoplasia) | 7,56 |
| Hepatocyte iron deficiency after induced siderosis | Similar to foci and carcinomas | 21 |
| Enzyme patterns | | |
|   Gamma-glutamyl transpeptidase | Similar to foci and carcinomas | 22,24,57 |
|   Glucose-6-phosphatase | Similar to foci and carcinomas | 27 |
| Appearance in noncirrhotic liver | Nondependence on chirrhosis or chronic hepatic damage | 10,11,60,68 |
| Alphafetoprotein production | Similar to carcinomas | 4 |
| Foci of carcinoma within adenoma | Progression to carcinoma | 12,46,55,60 |

importance in identifying carcinogens as to type or that classifying carcinogens as to mechanism of action is artificial and unproven, based merely on the lack of information about the mechanisms of carcinogenesis.

The induction of unique tumor types, even in one strain of mouse (B6C3F1), allows one to postulate the mechanisms of induction of these tumor types. According to the proposed carcinogen classification of Weisburger and Williams (45), one may hypothesize that (1) genotoxic hepatocarcinogens "induce" unique tumors or tumors that phenotypically and biologically resemble those in controls and (2) carcinogens acting by an epigenetic (nongenotoxic or unknown) mechanism would "induce" only tumors that morphologically and biologically resemble those in controls, since this type of carcinogen would simply increase the incidence of the

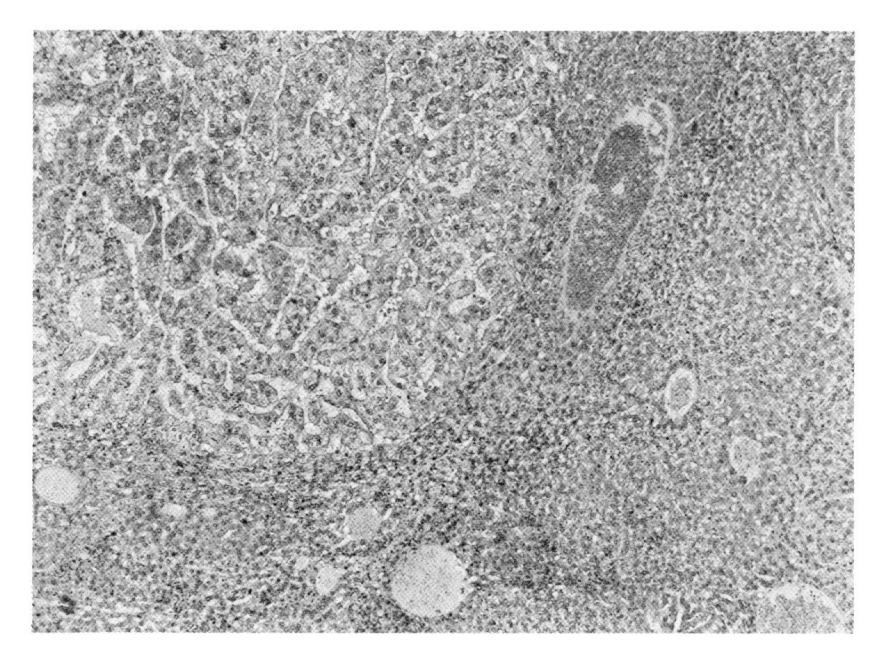

FIGURE 12. Focus of prominent trabecular formation (focal trabecular carcinoma) within a portion of a hepatocellular adenoma in a B6C3F1 mouse, X54.

TABLE 5. Comparative Aspects of the Metastatic Rate of Hepatocellular Tumors

| Species | Metastatic rate (%) Natural tumors | Induced tumors | Selected references |
|---|---|---|---|
| Mouse | 10-40 | 0-70 | 13,19,41,60 |
| Rat | 0 | 0-100 | 61,68 |
| Monkey | 0 | 0-30 | 38,62 |
| Human | 0-50 | N/A | 64,67 |

N/A, not applicable.

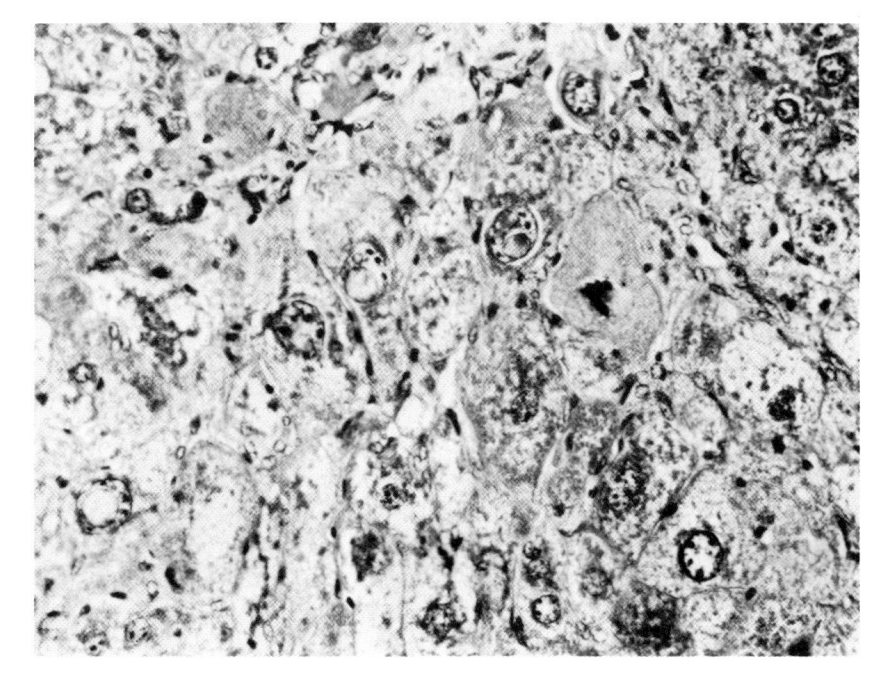

FIGURE 13. Mitotic figure in a chronically damaged liver of a mouse given nitrosodiethanolamine for 6 months. Note cytomegaly of other hepatocytes, X330.

background type of tumors found in controls. Both hypotheses appear to be wrong, since, exceptions to both are noted in Table 6. Then, what can be the cause(s) of specific morphologic types of liver tumors in mice receiving carcinogens? There may be three types of induction phenomena responsible for unique tumors, including (1) phenotypic expression of a genotoxic change (mutation), (2) chemical carcinogen effect on neoplastic hepatocytes as shown by induction of agranular endoplasmic reticulum and gamma-glutamyl transpeptidase by phenobarbital (59,70) and induction of peroxisomes by clofibrate and other peroxisome proliferators (30,31), and (3) enhancement of growth (promotion) of specific naturally occurring liver tumors. If epigenetic carcinogens induce unique tumors, they may do so by the second phenomenon. Eosinophilic tumors induced by nongenotoxic carcinogens may result from an effect on mitochondria and mitochondrial DNA (36,71,72).

OTHER HEPATIC TUMORS IN MICE

A variety of other nonhepatocellular primary hepatic tumors are found in mice (12,73). Comparative aspects of these tumors are shown in Table 7. Hepatoblastomas are controversial tumors in aged mice (74). Some authors have considered them to be cholangiocarcinomas (12). Histocytic lymphomas may arise in the

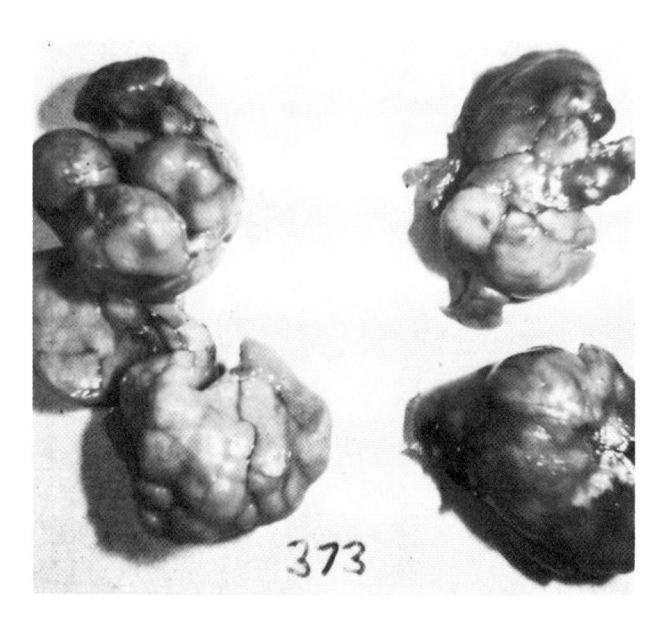

FIGURE 14. Multiple coalescing masses of hepatocellular adenomas and carcinomas in livers of mice given 2-aminothiazole.

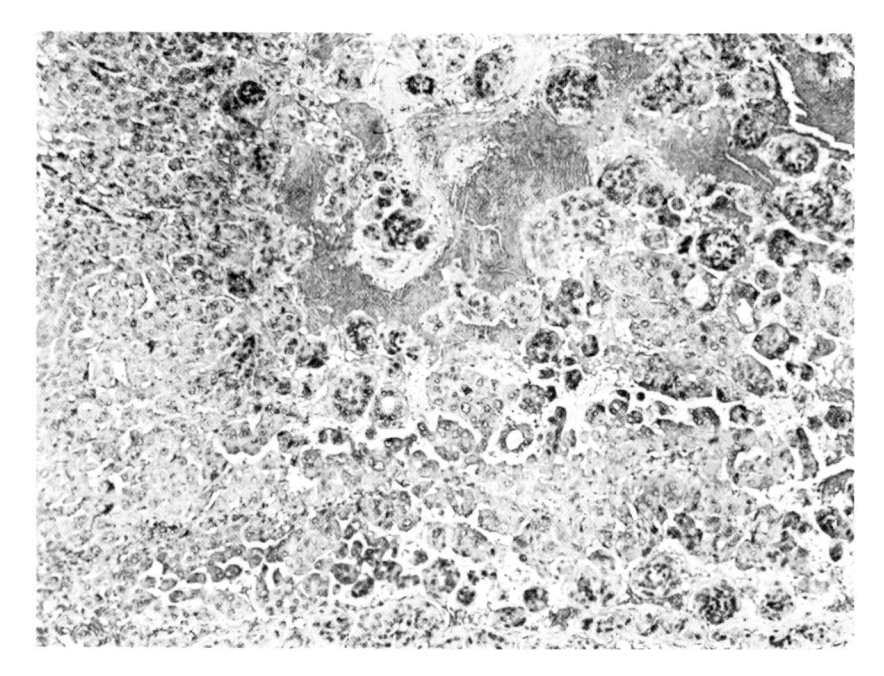

FIGURE 15. Trabecular hepatocellular carcinoma in a B6C3F1 mouse. Note thrombosis of vascular spaces, X180.

TABLE 6. Chemicals Inducing Eosinophilic or Basophilic Hepatocellular Tumors in Mice

| Type of tumor | Type of carcinogen[a] | Chemical | Metastatic rate to lung (%) | Selected references |
|---|---|---|---|---|
| Basophilic | Epigenetic (nongenotoxic) | Chloroform | 0-5 | [b] |
| | | Tetrachlorvinphos | 0-5 | [b] |
| | | Di(2-ethylhexyl)phthalate | 30 | [b] |
| | Genotoxic | Toxophene | 0 | [b] |
| | | Diethylnitrosamine | 31 | 19,41 |
| | | 2-Acetylaminofluorene | 0-20 | 12 |
| | | Benzidine | 5-68 | 10,11,13 |
| Eosinophilic | Epigenetic (nongenotoxic) | Phenobarbital | 0 | 5,7,70 |
| | | Nitrofen | 10-25 | 15 |
| | | α-Benzene hexachloride | 0 | 16 |
| | | Chlordane | 0-5 | [b] |
| | | Dieldrin | 0-10 | 36 |
| | Genotoxic | 2-Nitro-p-phenylenediamine | 0 | [b] |
| | | 4-Chloro-m-phenylenediamine | 0 | [b] |

[a] Based on a classification proposed by Weisburger and Williams (72).
[b] National Cancer Institute of National Toxicology Program, Carcinogenesis. Technical Report Series, available from the National Technical Information Service, S285 Port Royal Rd., Springfield, VA 22161.

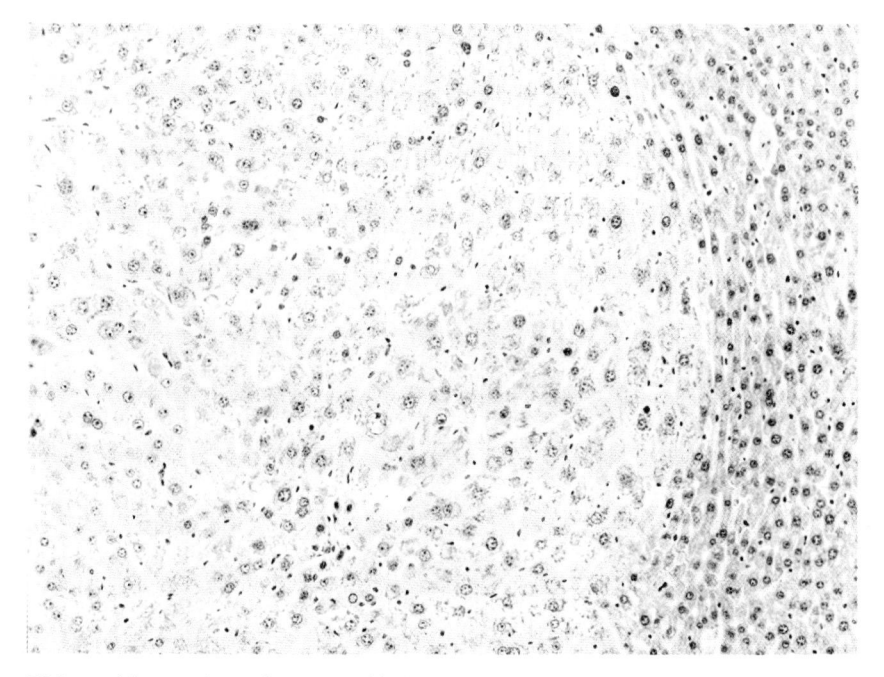

FIGURE 16. Unique hepatocellular adenoma induced by toxophene in a B6C3F1 mouse. The tumor is composed of hepatocytes with pale basophilic cytoplasm and prominent clusters of granular endoplasmic reticulum, X130.

livers of male mice (75). Such tumors may arise from either Kupffer's cells or true histocytes. The majority of mouse liver carcinogens induce hepatocellular tumors. Occasionally, chemicals induce these and other types of liver tumors: for example, nitrofen-induced hepatocellular tumors and the controversial hepatoblastoma (15). Few mouse liver carcinogens induce nonhepatocellular tumors only.

SUMMARY OF COMPARATIVE ASPECTS

A review of the histogenesis, morphology, and biology of hepatocellular tumors of mice has revealed characteristics similar to those of rats and other species. Although differences may be noted for some characteristics, the similarities in other cases are striking. Similarities to other epithelial tumors of mice and other species are also evident. While these comparative findings do not lessen the significance of mouse liver tumors in carcinogenesis bioassays, other characteristics of the mouse liver may do so. It has been suggested that the mouse liver may be more sensitive to the carcinogenic effects of liver toxins (6) compared to the livers of other species. In addition, the high doses of chemicals used in bioassays may permit artificial situations not comparable

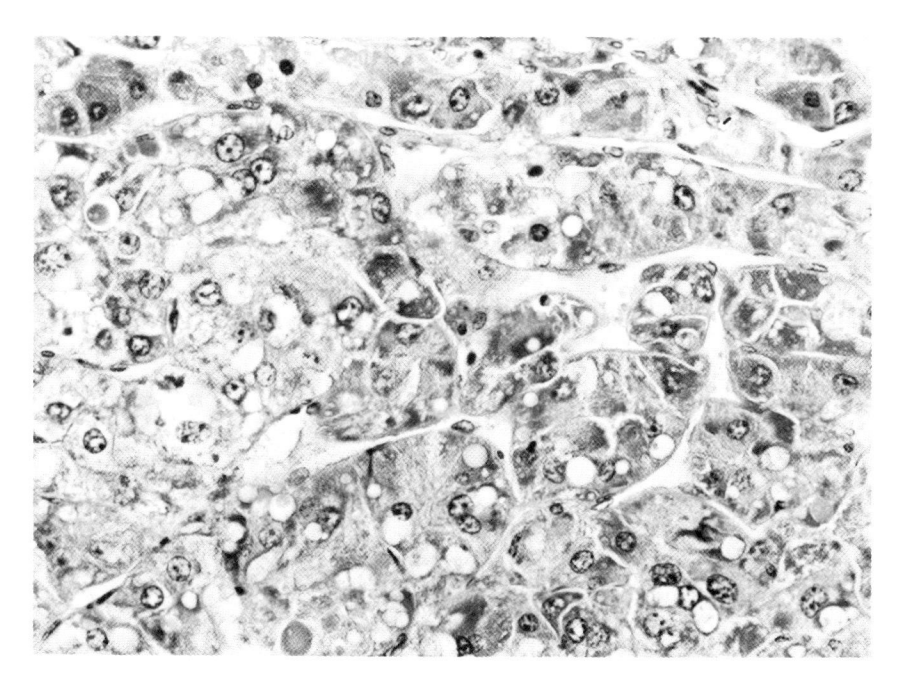

FIGURE 17. Unique hepatocellular carcinoma induced by tetrachlorvinphos in a B6C3F1 mouse. The tumor is trabecular and composed of vacuolated hepatocytes with occasional intracytoplasmic inclusions, X330.

to those in humans. Many strains of mice have high natural rates of liver tumor occurrences. While these arguments may have some significance for extrapolating mouse data to human risk assessment, an important question concerning bioassay findings arise when mouse liver tumors are the only endpoint in a bioassay test. If a chemical were found to cause only liver tumors in mice and no tumors of any type in other species, including humans, the significance of tumors might justifiably be questioned. However, DDT was originally found to cause only mouse liver tumors, (60) but several years later liver tumors were induced in rats. Limited experimental data can be no excuse for drawing incorrect conclusions. Most chemicals, demonstrated to be carcinogenic in humans, have been found in workers exposed to high toxic doses, a situation similar to that in rodent bioassays. Further laboratory and epidemiological studies on the chemicals that have only been found to be carcinogenic for mouse liver may reveal surprising findings.

REFERENCES

1. Armuth V, Berenblum I: Systemic promoting action of phorbol in liver and lung carcinogenesis in AKR mice. Cancer 32:2259-2262, 1972.

TABLE 7. Comparative Aspects of Other Liver Tumors

| Species | Hemangiosarcoma | Hepatoblastoma | Cholangiocarcinoma | Histiocytic sarcoma | Selected references |
|---------|-----------------|----------------|--------------------|--------------------|--------------------|
| Mouse | Common | Rare; old mice | Rare | Common | 12,73-75 |
| Rat | Rare | Rare, genetic | Rare | CD Rat | 58 |
| Monkey | Rare | Rare | Aflatoxin | Rare | 62 |
| Human | Vinyl chloride thorotrast | Children | Clonorchis | Rare | 64 |

2. Bannasch P, Mayer D, Venske G: Pranatale induktion von hepatocellularen glykogenspeicherarealen und tumoren bei mausen durch athylnitrosoharnstoff. Virchows Arch (Zellpathol) 30:143-160, 1979.

3. Becker FF: Characterization of sponaneous and chemically induced mouse liver tumors. Chap. 10 in this volume.

4. Becker FF, Sell S: γ-Fetoprotein levels and hepatic alterations during chemical carcinogenesis in C57BL/6N mice. Cancer Res 39:3491-3494, 1979.

5. Butler, WH, Hempsall V: Histochemical observations on nodules induced in the mouse liver by phenobarbitone. J Pathol 125:155-161, 1978.

6. Butler WH, Jones G: Pathological and toxicological data on chlorinated pesticides and phenobarbital. Ecotoxicol Environ Safety 1:503-509, 1978.

7. Butler WH, Newberne PM, editors: Mouse Hepatic Neoplasia, p. 195. Amsterdam: Elsevier, 1975.

8. Chu KC, Cueto C Jr., Ward JM: Factors in the evaluation of 200 National Cancer Institute carcinogen bioassays. J Toxicol Environ Health 8:251-280, 1981.

9. Essigman EM, Newberne PM: Enzymatic alterations in mouse hepatic nodules induced by a chlorinated hydrocarbon pesticide. Cancer Res 41:2823-2831, 1981.

10. Frith CH, Baetcke KP, Nelson CJ, Schieferstein G: Sequential morphogenesis of liver tumors in mice given benzidine dihydrochloride. Eur J Cancer 16:1205-1216, 1980.

11. Frith CH, Dooley K: Hepatic cytologic and neoplastic changes in mice given benzidine dihydrochloride. J Natl Cancer Inst 56:679-682, 1976.

12. Frith CH, Ward JM: A morphologic classification of proliferative and neoplastic hepatic lesions in mice. J Environ Pathol Toxicol 3:329-351, 1979.

13. Frith CH, Baetcke KP, Nelson CJ, Schieferstein G: Correlation of liver tumor morphology and weight to incidence of pulmonary metastases in the mouse. Toxicol Lett 7:113-118, 1980.

14. Goldfarb S, Pugh TD, Cripps DJ: Increased alkaline phosphatase activity - A positive histochemical marker for griseofulvin-induced mouse hepatocellular nodules. J Natl Cancer Inst. 64:1427-1433, 1980.

15. Hoover KL, Stinson SF, Ward JM: Histopathologic differences between liver tumors in untreated (C57BL/6 x C3H)F1 (B6C3F1)

mice and nitrofen-fed mice. J Natl Cancer Inst 65:937-948, 1980.

16. Ito N, Hananouchi M, Sugihara S, Shirai T, Tsuda H, Fukushima S, Nagasaki H: Reversibility and irreversibility of liver tumors in mice induced by the α-isomer of 1,2,3,4,5,6-hexachlorocyclohexane. Cancer Res 36:2227-2234, 1976.

17. Jalanko H, Ruoslahti E: Differential expression of γ-fetoprotein and γ-glutamyltranspeptidase in chemical and spontaneous hepatocarcinogenesis. Cancer Res 39:3495-3501, 1979.

18. Kimbrough RD, Linder RE: Induction of adenofibrosis and hepatomas of the liver in BALB/c mice by polychlorinated biphenyls (Aroclor 1254). J Natl Cancer Inst 53:547-552, 1974.

19. Kyriazis AP, Vessilinovitch SD: Transplantability and biological behavior of mouse liver tumors induced by ethylnitrosourea. Cancer Res 33:332-338, 1973.

20. Lipsky MM, Tanner DC, Hinton DE, Trump BE: Reversibility, persistence, and progression of safrole-induced mouse liver lesions following cessation of exposure. Chap. 11 in this volume.

21. Lipsky MM, Hinton DE, Goldblatt PJ, Klaunig JE, Trump BF: Iron negative foci and nodules in safrole-exposed mouse liver made siderotic by iron-dextran injection. Path Res Pract 164:178-185, 1979.

22. Lipsky MM, Hinton DE, Klaunig JE, Goldblatt PJ, Trump BF: Gamma-glutamyl transpeptidase in safrole-induced, presumptive premalignant mouse hepatocytes. Carcinogenesis 1:151-156, 1980.

23. Lipsky MM, Hinton DE, Klaunig JE, Trump BF: Biology of hepatocellular neoplasia in the mouse. I. Histogenesis of safrole-induced hepatocellular carcinoma. J Natl Cancer Inst 67:365-376, 1981.

24. Lipsky MM, Hinton DE, Klaunig JE, Goldblatt PJ, Trump BF: Biology of hepatocellular neoplasia in the mouse. II. Sequential enzyme histochemical analysis of BALB/c mouse liver during safrole-induced carcinogenesis. J Natl Cancer Inst 67:377-392, 1981.

25. Lipsky MM, Hinton DE, Klaunig JE, Goldblatt PJ, Trump BF: Biology of hepatocellular neoplasia in the mouse. III. Electron microscopy of safrole-induced hepatocellular adenomas and hepatocellular carcinomas. J Natl Cancer Inst 67:393-405, 1981.

26. Meierhenry EF, Ruebner BH, Gershwin ME, Heieh BS, French SW: Mallory body formation in hepatic nodules of mice ingesting dieldrin. Lab Invest 44:392-396, 1981.

27. Moore MR, Drinkwater NR, Miller EC, Miller JA, Pitot HC: Quantitative analysis of the time dependent development of glucose-6-phophatase-deficient foci in the livers of mice treated neonatally with diethylnitrosamine. Cancer Res 41:1585-1593, 1981.

28. Ohmori T, Rice JM, Williams GM: Histochemical characteristics of spontaneous and chemically induced hepatocellular neoplasms in mice. Histochem J 13:85-99, 1981.

29. Peraino C, Fry RJM, Staffeldt E: Enhancement of spontaneous hepatic tumorigenesis in C3H mice by dietary phenobarbital. J Natl Cancer Inst 51:1349-1350, 1973.

30. Reddy JK, Rao MS, Azarnoff DL, Sell S: Mitogenic and carcinogenic effects of a hypolipidemic peroxisome proliferator (4-chloro-6-(2,3-xylidino)-2-pyrimidinylthio) acetic acid (Wy-14,643) in rat and mouse liver. Cancer Res 39:152-161, 1979.

31. Reddy JK, Rao MS, Moody DE: Hepatocellular carcinomas in acatalasemic mice treated with nafenopin, a hypolipidemic peroxisome proliferator. Cancer Res 36:1211-1217, 1976.

32. Reuber MD: Carcinomas and other lesions of the liver in mice ingesting organochlorine pesticides. Clin Toxicol 13:231-256, 1978.

33. Reuber MD: Histogenesis of hyperplasia and carcinomas of the liver arising around central veins in mice ingesting chlorinated hydrocarbons. Pathol Microbiol 43:287-298, 1975.

34. Reuber MD, Ward JM: Histopathology of liver carcinomas in (C57BL/6N x C3H/HeN)F1 mice ingesting chlordane. J Natl Cancer Inst 63:89-92, 1979.

35. Reznik G, Ward JM: Carcinogenicity of the hair-dye component 2-nitro-p-phenylenediamine: Induction of eosinophilic hepatocellular neoplasms in female B6C3F1 mice. Food Cosmet Toxicol 17:493-500, 1979.

36. Reubner BH, Gershwin ME, French SW, Meierhenry E, Dunn P, Hsieh LS: Mouse hepatic neoplasia: Differences among strains and carcinogens. Chap. 9 in this volume.

37. Reubner BH: Ultrastructure of spontaneous neoplasms induced by diethylnitrosoamine and dieldrin in the C3H mouse. J Environ Pathol Toxicol 4:237-254, 1981.

38. Reubner BH, Michas C, Kanayama R, Bannasch P: Sequential hepatic histologic and histochemical changes produced by

diethylnitrosamine in the rhesus monkey. J Natl Cancer Inst 57:1261-1267, 1976.

39. Stinson SF, Hoover KL, Ward JM: Quantitation of differences between spontaneous and induced liver tumors in mice with an automated image analyzer. Cancer Lett 14:143-150, 1981.

40. Tomatis L, Turusov V, Day N, Charles RT: The effect of long-term exposure to DDT on CF-1 mice. Int J Cancer 10:489-506, 1972.

41. Vesselinovitch SD, Mihailovich N, Rao KVN: Morphology and metastatic nature of induced hepatic nodular lesions in C57BL x C3H F1 mice. Cancer Res 38:2003-2010, 1978.

42. Walker AIT, Thorpe E, Stevenson DE: The toxicology of dieldrin (HEOD). I. Long-term oral toxicity studies in mice. Food Cosmet Toxicol 11: 415-432, 1972.

43. Ward JM, Griesemer RA, Weisburger EK: The mouse liver tumor as an endpoint in carcinogenesis tests. Toxicol Appl Pharmacol 51:389-397, 1979.

44. Ward JM, Bernal E, Buratto B, Goodman DG, Strandberg JD, Schueler R: Histopathology of neoplastic and non-neoplastic lesions in mice fed diets containing tetrachlorvinphos. J Natl Cancer Inst 63:111-118, 1979.

45. Weisburger JH, Williams GM: Carcinogen testing: Current problems and new approaches. Science 214:401-107, 1981.

46. Andervont HB, Dunn TB: Transplantation of spontaneous and induced hepatomas in inbred mice. J Natl Cancer Inst 13:455-503, 1952.

47. Becker FF, Stillman D, Sell S: Serum-fetoprotein in a mouse strain (C3H-AvyfB) with spontaneous hepatocellular carcinomas. Cancer Res 37:870-872, 1977.

48. Essner E: Ultrasturcture of spontaneous hyperplastic nodules in mouse liver. Cancer Res 27:2137-2152, 1967.

49. Heston WE, Vlahakis G: Factors in the causation of spontaneous hepatomas in mice. J Natl Cancer Inst 37:839-843, 1966.

50. Hruban Z, Kirsten WH, Slesers A: Fine structure of spontaneous hepatic tumors of male C3H/fGs mice. Lab Invest 15:576-588, 1966.

51. Reuber MD: Morphologic and biologic correlation of hyperplastic lesions occurring "spontaneously" in C3H x Y hybrid mice. Br J Cancer 25:538-543, 1971.

52. Sharp JG, Riches AC, Littlewood V, Thomas DB: The incidence, pathology and transplantation of hepatomas in CBA mice. J Pathol 119:211-220, 1976.

53. Tarone RE, Chu KC, Ward JM: Variability in the rates of some common naturally occurring tumors in F344 rats and B6C3F1 mice. J Natl Cancer Inst 66:1175-1181, 1981.

54. Ward JM: Background data and variations in tumor rates of control rats and mice. Prog Exp Tumor Res 26:241-258, 1983.

55. Ward JM, Vlahakis G: Evaluation of hepatocellular neoplasms in mice. J Natl Cancer Inst 61:807-811, 1978.

56. Williams GM, Hirota N, Rice JM: The resistance of spontaneous mouse hepatocellular neoplasms to iron accumulation during rapid iron loading by parenteral administration and their transplantability. Am J Pathol 94:65-74, 1979.

57. Williams GM, Ohmori T, Katayama S, Rice JM: Alteration by phenobarbital of membrane-associated enzymes including gamma-glutamyl transpeptidase in mouse liver neoplasms. Carcinogenesis 1:813-818, 1980.

58. Squire RA, Levitt M: Report of a workshop on classification of specific hepatocellular lesions in rats. Cancer Res 35:3214-3215, 1975.

59. Stewart HL: Comparative aspects of certain cancers. In: Cancer - A Comprehensive Treatise, edited by FF Becker, vol. 4, p. 320-329. New York: Plenum, 1975.

60. Ward JM: Morphology of hepatocellular neoplasms in B6C3F1 mice. Cancer Lett 9:319-325, 1980.

61. Williams GM: The pathogenesis of rat liver cancer caused by chemical carcinogens. Biochim Biophys Acta 605:167-189, 1980.

62. Sieber SM, Correa P, Dalgard DW, Adamson RH: Induction of osteogenic sarcomas and tumors of the hepatobiliary system in nonhuman primates with aflatoxin $B_1$. Cancer Res 39:4545-4554, 1979.

63. Anthony PP: Precancerous changes in the human liver. J Toxicol Environ Health 5:301-313, 1979.

64. Cameron HM, Warwick GP: Primary cancer of the liver in Kenyan children. Br J Cancer 36:793-803, 1977.

65. Cohen C, Berson SD, Geddes EW: Liver cell dysplasia: Association with hepatocellular carcinoma, cirrhosis and hepatitis B antigen carrier status. Cancer 44:1671-1676, 1979.

66. Parker P, Burr I, Slonim A, Ghishan FK, Greene H: Regression of hepatic adenomas in type Ia glycogen storage disease with dietary therapy. Gastroenterology 81:534-536, 1981.

67. Uchida T, Miyata H, Shikata T: Human hepatocellular carcinoma and putative precancerous disorders. Arch Pathol Lab Med 105:180-186, 1981.

68. Ward JM, Reznik G: Refinements of rodent pathology and the pathologist's contribution to evaluation of carcinogenesis. Prog Exp Tumor Res 26:266-291, 1983.

69. Stewart HL, Snell KC, Dunham LJ, Schlyen SM: Transplantable and transmissible tumors of animals. Atlas of tumor pathology. Section XII - Fasc. 40, p. 378. Washington, D.C.: Armed Forces Institute of Pathology, 1959.

70. Ward JM, Rice JM, Creasia D, Lynch P, Riggs C: Dissimilar patterns of promotion by di(2-ethylhexyl)phthalate and phenobarbital of hepatocellular neoplasia initiated by diethylnitrosamine in B6C3F1 mice. Carcinogenesis 4:1021-1029, 1983.

71. Reznik-Schuller HM, Lijinsky W: Morphology of early changes in liver carcinogenesis induced by methapyrilene. Arch Toxicol 49:79-83, 1981.

72. Neubert D, Hopfenmuller W, Fuchs G: Manifestation of carcinogenesis as a stochastic process on the basis of an altered mitochondrial genome. Arch Toxicol 48:89-125, 1981.

73. Turusov VS, Takayama S: Tumours of the liver. In: Pathology of Tumours in Laboratory Animals, volume II, Tumours of the Mouse, pp. 193-233. Lyon: IARC, 1979.

74. Turusov VS, Deringer MK, Dunn TB, Stewart HL: Malignant mouse-liver tumors resembling human hepatoblastomas. J Natl Cancer Inst 51:1689-1695, 1973.

75. Frith CH, Davis TM, Zolotor LA, Townsend JW: Histocytic lymphoma in the mouse. Leuk Res 4:651-662, 1980.

Addendum - Some recent references on the histogenesis of mouse liver tumors include Koen et al., Amer J Pathol 112:89-100, 1983 and Vessilinovitch et al., Cancer Research 43:4253-4259, 1983.

Chapter 2

# Occurrence of Mouse Liver Neoplasms in the National Cancer Institute Bioassays

**Thomas E. Hamm, Jr.**

INTRODUCTION

This chapter discusses the occurrence of liver tumors in the mouse in 2-year bioassays and the importance of these tumors in determining the results of those bioassays. The National Cancer Institute (NCI) bioassays reported between 1977 and 1980 have been selected, since they have several advantages. First, in all of the reported NCI bioassays, the B6C3F1 mouse was used. This mouse was produced by crossing C57BL6/N females with C3H/HeN MTV- males. The N refers to the National Institutes of Health (NIH) strains, and the mice were produced by commercial animal producers using parental lines from the NIH colonies. This minimized the effect of genetic differences between groups of mice in different bioassays. However, no genetic monitoring of animals shipped to bioassay laboratories was done. Second, both sexes were used so differences between the sexes would not be overlooked. Third, husbandry procedures were standardized to minimize environmental variability within and between laboratories (1,2). For example, in most bioassays the animals were housed, 5 animals per cage, in a standard-size cage developed for the bioassay program. Bedding was heat-treated hardwood chips. Feed was a closed-formula standard rodent diet from one of two suppliers. Fourth, necropsy at a standardized 103-110 weeks helped minimize variable tumor incidence caused by accelerated tumor formation in older animals. Finally, a large number of bioassays are available for evaluation and comparison.

The experimental design for the NCI bioassay is summarized in Table 1. More information on the evaluation and design of these bioassays can be found in Chu et al. (1) and Sontag et al. (2), respectively. While both rats and mice were used in most bioassays, the mice were always B6C3F1 hybrids and the rats were either Fischer 344 (F-344), Osborne Mendel (OM), or Sprague-Dawley. Animals from both sexes of each species were divided into three groups: control, low dose and high dose. Usually each group consisted of 50 animals, but some earlier bioassays used only 20 control animals. The major route of chemical administration was

TABLE 1. NCI Bioassay Experimental Design Summary

| Group[a] | Initial Group size | Dosage | Observation period |
|---|---|---|---|
| Control | 20-50 | Vehicle | 104-110 weeks |
| Low dose | 50 | ½ MTD | 104-110 weeks |
| High dose | 50 | MTD | 104-110 weeks |

[a]One of each group for each sex of each species.

MTD, maximum tolerated dose

dosed feed. The high dose was selected to be the maximum tolerated dose (MTD), based on the results of short-term tests including a 90-day study. The MTD was defined as the highest dose that could be given that would not cause a significant decrease in survival from effects other than carcinogenicity. The ideal MTD was to cause no more than a 10% reduction in body weight gain and not exceed 5% of the diet, to avoid affecting the nutritive balance of the diet. The test period was 104-110 weeks, after which all survivors were necropsied and examined for lesions.

This review was conducted using data in the original reports* and the paper in IARC scientific publication no. 27 by Griesemer and Cueto (3). The paper by Griesemer and Cueto was particularly useful and should be consulted for additional information about bioassays not included in this presentation, for information about the incidence of nonhepatic tumor types and for the interpretation and classification of these bioassay experiments.

There were 192 bioassays available for review. Approximately half (94 bioassays) were not considered further, since there was no evidence (66 bioassays) or only equivocal evidence (28 bioassays) of any type of chemically induced tumor in either species. This left 98 studies with evidence of carcinogenicity in at least one sex of one species. Three of these studies were eliminated from consideration because only rats were used. Ten more were not considered because they were conducted using B6C3F1 mice and Sprague-Dawley rats, usually by the intraperitoneal injection of the test chemical and did not provide a large enough group of

---

*Reports may be obtained from the National Technical Information Service, U.S. Department of Labor, 5285 Port Royal Rd., Springfield, VA 22161

similar experiments to make useful comparisons. Thus, 85 bioassays were used for this review.

The morphology and histopathologic characteristics used to diagnose the mouse liver tumor in this series of bioassays have been discussed by Dr. J. M. Ward (Chapter 1). In all bioassays, adenomas and adenocarcinomas of the liver were considered together for statistical evaluation. The pathology working group headed by Dr. Ward was responsible for assurance that the tumors reported to the bioassay by contractors to the bioassay were diagnosed according to standard criteria (4).

Tarone et al. (5), using data from these bioassays, have shown significant interlaboratory variation in the incidence of liver tumors in control B6C3F1 mice. Laboratory 1 had a mean of 40.1% (24-58) for males and 9.7% (2-21) for females, while the corresponding values for the other laboratories were: laboratory 2, 31.3% (17-39), 4.6% (2-10); laboratory 3, 25% (16-39), 7.3% (0-13); laboratory 4, 32.2% (15-55), 5.1% (0-21); and laboratory 5, 27.4% (7-45), 4.8% (0-17). Data on all neoplastic lesions found in control animals from these bioassays have been published for B6C3F1 mice (6), F-344 rats (7) and OM rats (8).

Tables 2 and 3 list the chemicals that induced a statistically increased incidence of tumors in at least one sex of one species. Table 2 lists those bioassays in which OM rats were used, and Table 3 lists those in which F-344 rats were used. The comparison is between the occurrence of liver tumors and/or all other tumors, combined by sex by species. A $\pm$ means that the data was suggestive, but not conclusive, that tumors were produced by the chemicals. When a $\pm$ occurred, it was considered a minus for the remaining tables. The report numbers are listed for those requiring more information on specific chemicals.

Of the 85 bioassays positive for any tumor type in at least one sex of one species, the B6C3F1 mouse had only liver tumors in 36 of those bioassays. The mice had tumors in both the liver and other sites in 14 bioassays, and only nonhepatic tumors in 15 bioassays. There was no increase in any tumors in 20 bioassays. The F-344 rat was used in 60 of the 85 bioassays and had only liver tumors in 8, both liver and nonhepatic tumors in 8 bioassays, only nonhepatic tumors in 33, and no statistical increase in any tumors in 11. The OM rat was used in 25 of the 85 bioassays. This rat had only liver tumors in 1 bioassay, both liver tumors and nonhepatic tumors in 2 bioassays, only nonhepatic tumors in 6 bioassays, and there was no increase in any tumors in 16 bioassays. Liver tumors were more common in the B6C3F1 mouse than in the F-344 rat or OM rat and were the most frequently diagnosed tumor in the mouse (Table 4).

Except for nitrofen (17 in Table 2 and 42 in Table 3), the chemicals tested in OM rats differed from those tested in F-344 rats, so direct comparisons between the responses of the two rat types to chemicals cannot be made. (The OM rats had pancreatic

TABLE 2. Chemicals That Induced Tumors in B6C3F1 Mice or OM Rats

| Chemical | B6C3F1 mice Liver M F | B6C3F1 mice Other M F | OM rats Liver M F | OM rats Other M F | Report number |
|---|---|---|---|---|---|
| 1. Aldrin | + − | − − | − − | + + | 21 |
| 2. Captan | − − | + +/− | − − | − − | 15 |
| 3. Chloramben | +/− − | − − | − − | − − | 25 |
| 4. Chlordane | +/− + | − − | − − | +/− + | 8 |
| 5. Chlordecone | + + | − − | + + | − − | None |
| 6. Chlorobenzilate | + + | − − | − − | − − | 75 |
| 7. Chloroform | + + | − − | − − | + − | None |
| 8. Chlorothalonil | − − | − − | − − | + + | 41 |
| 9. p,p'-DDE | + + | − − | − − | − − | 131 |
| 10. Dibromochloropropane | − − | + + | − − | + + | 28 |
| 11. 1,2-Dibromoethane | − − | + + | − + | + + | 86 |
| 12. 1,2-Dichloroethane | − − | + + | − − | + + | 55 |
| 13. Dicofol | + − | − − | − − | − − | 90 |
| 14. 1,4-Dioxane | + + | − − | − + | + + | 80 |
| 15. Heptachlor | + + | − − | − − | − − | 9 |
| 16. Hexachloroethane | + + | − − | − − | − − | 68 |
| 17. Nitrofen | − − | + − | − − | − + | 26 |
| 18. Sulfallate | − − | + + | − − | + + | 115 |
| 19. 1,1,2,2-Tetrachloroethane | + + | − − | +/− − | − − | 27 |
| 20. Tetrachloroethylene | + + | − − | − − | − − | 13 |
| 21. Tetrachlorvinphos | +/− + | − − | − − | − +/− | 33 |
| 22. Toxaphene | + + | − − | − − | +/− +/− | 37 |
| 23. 1,1,2-Trichloroethane | + + | +/− + | − − | − − | 74 |

TABLE 2. Chemicals That Induced Tumors in B6C3F1 Mice or OM Rats (continued)

|  | B6C3F1 mice | | OM rats | | |
|---|---|---|---|---|---|
| Chemical | Liver<br>M F | Other<br>M F | Liver<br>M F | Other<br>M F | Report<br>Number |
| 24. Trichloroethylene | + + | − − | − − | − − | 2 |
| 25. Trifluralin | − + | − − | − − | − − | 34 |

+ = Significantly more tumors than control animals.
− = Tumor incidence no greater than controls.
+ = Tumor incidence suggestive of an increase above controls.
─

TABLE 3. Chemicals That Induced Tumors in B6C3F1 Mice or F-344 Rats

| Chemical | B6C3F1 mice Liver M F | B6C3F1 mice Other M F | F-344 rats Liver M F | F-344 rats Other M F | Report number |
|---|---|---|---|---|---|
| 1. 2-Aminoanthraquinone | + + | − + | + − | − − | 144 |
| 2. 3-Amino-4-ethoxyacetanilide | − − | + − | − − | − − | 112 |
| 3. 3-Amino-9-ethylcarbazole HCl | + + | − − | + + | + + | 93 |
| 4. 1-Amino-2-methylanthraquinone | − + | +/− − | + + | + − | 111 |
| 5. 4-Amino-2-nitrophenol | − − | − − | − − | + +/− | 94 |
| 6. 2-Amino-5-nitrothiazole | − − | − − | − − | + + | 53 |
| 7. Aniline HCl | − − | + + | − − | + + | 130 |
| 8. o-Anisidine HCl | − − | + + | − − | + + | 89 |
| 9. Azobenzene | − − | − − | − − | + − | 154 |
| 10. 3-(Chloromethyl)pyridine HCl | − − | + + | − − | + − | 95 |
| 11. 4-Chloro-m-phenylenediamine | − + | − − | − − | + − | 85 |
| 12. 4-Chloro-o-phenylenediamine | + + | − − | − − | + + | 63 |
| 13. 5-Chloro-o-toluidine | + + | + + | − − | − − | 187 |
| 14. 4-Chloro-o-toluidine HCl | − − | + + | − − | − − | 165 |
| 15. Cinnamyl anthranilate | + + | − − | − − | − − | 196 |
| 16. C. I. vat yellow 4 | − − | + − | − − | − − | 134 |
| 17. m-Cresidine | − − | + + | − − | + + | 105 |
| 18. p-Cresidine | − + | + + | − − | + + | 142 |
| 19. Cupferron | − + | + + | + + | + + | 100 |
| 20. Diaminozide | +/− − | − − | − − | − + | 83 |
| 21. Dapsone | − − | − − | − − | + − | 20 |
| 22. 2,4-Diaminoanisole | − − | + + | − − | + + | 84 |
| 23. 2,4-Diaminotoluene | − + | − +/− | + + | + + | 162 |
| 24. N,N'-Diethylthiourea | − − | − − | − − | + + | 149 |

TABLE 3. Chemicals That Induced Tumors in B6C3F1 Mice or F-344 Rats (continued)

| | | B6C3F1 mice | | | F-344 rats | | |
| | | Liver M F | Other M F | Liver M F | Other M F | Report number |
|---|---|---|---|---|---|---|
| 25. | 3,3'-Dimethoxybenzidine-4,4'-diisocyanate | – – | – – | – – | + + | 128 |
| 26. | 2,4-Dinitrotoluene | – – | – – | – – | + + | 54 |
| 27. | Direct black 38 | – – | – – | + + | – – | 108 |
| 28. | Direct blue 6 | – – | – – | + + | – – | 108 |
| 29. | Direct brown 95 | – – | – – | – + | – – | 108 |
| 30. | Ethyl tellurac | – – | + + | – – | +/– – | 152 |
| 31. | Hydrazobenzene | – + | – – | – – | +/+ + | 92 |
| 32. | 4,4'-Methylene-bis(N,N-dimethyl)-benzenamine | +/– + | – – | + + | + + | 186 |
| 33. | Michler's ketone | – + | + – | – – | – – | 181 |
| 34. | 1,5-Naphthalenediamine | – + | + + | – – | – + | 143 |
| 35. | Nithiazide | + +/– | – – | – – | – + | 146 |
| 36. | Nitrolotriacetic acid | – – | + +/– | – – | + + | 6 |
| 37. | Nitrolotriacetic acid, trisodium salt | – – | +/– – | – – | + + | 6 |
| 38. | 5-Nitroacenaphthene | – + | – + | – – | + + | 118 |
| 39. | 3-Nitro-p-acetophenetide | + – | – – | – – | – – | 133 |
| 40. | 5-Nitro-o-anisidine | +/+ + | – – | – – | + + | 127 |
| 41. | 6-Nitrobenzimidazole | +/+ + | – – | – – | – – | 117 |
| 42. | Nitrofen | – + | – – | – – | – – | 184 |
| 43. | 2-Nitro-p-phenylenediamine | – + | – – | – – | – – | 169 |
| 44. | 3-Nitropropionic acid | – – | – – | + – | +/– – | 52 |
| 45. | N-Nitrosodiphenylamine | – – | – – | – – | +/+ + | 164 |
| 46. | p-Nitrosodiphenylamine | + – | – – | + – | – – | 190 |
| 47. | 5-Nitro-o-toluidine | + + | + + | – – | – – | 107 |
| 48. | Phenazopyridine HCl | – + | – – | – – | + + | 99 |
| 49. | Piperonyl sulfoxide | + – | – – | – – | – – | 124 |
| 50. | Pivalolactone | – – | – – | – – | + + | 140 |

TABLE 3. Chemicals That Induced Tumors in B6C3F1 Mice or F-344 Rats (continued)

|  |  | B6C3F1 mice | | F-344 rats | | Report |
|---|---|---|---|---|---|---|
| | Chemical | Liver<br>M F | Other<br>M F | Liver<br>M F | Other<br>M F | number |
| 51. | p-Quinone dioxime | − − | − − | − − | + +<br>+ − | 179 |
| 52. | Reserpine | − − | + + | − − | + − | 193 |
| 53. | Selenium sulfide | − + | − + | + + | − − | 194 |
| 54. | 4,4'-Thiodianiline | + + | + + | + − | + + | 47 |
| 55. | o-Toluidine HCl | − + | + − | − − | + − | 153 |
| 56. | 2,4,6-Trichlorophenol | + + | − − | + + | + − | 155 |
| 57. | 2,4,5-Trimethylaniline | − + | − − | − − | − + | 160 |
| 58. | Trimethyl phosphate | − − | − + | − − | + − | 81 |
| 59. | Trimethylthiourea | − − | − − | − − | − + | 129 |
| 60. | Tris(2,3-dibromopropyl)phosphate | − + | + + | − − | + + | 76 |

+ = Significantly more tumors than control animals.
− = Tumor incidence no greater than controls.
+ = Tumor incidence suggestive of an increase above controls.
−

TABLE 4. Tumors in Rodents in NCI Bioassays[a]

| Tumor type | B6C3F1 mouse | F-344 rat | OM rat |
|---|---|---|---|
| Liver only | 36 | 8 | 1 |
| Liver and nonhepatic | 14 | 8 | 2 |
| Nonhepatic only | 15 | 33 | 6 |
| No tumors | 20 | 11 | 16 |
| Total | 85 | 60 | 25 |

[a] Bioassay with significantly increased tumors in at least one sex of one species.

tumors in females, and no tumors occurred in F-344 rats with nitrofen exposures.)

A comparison of tumor occurrence between the B6C3F1 mouse and the F-344 rat in the same bioassay is presented in Table 5. When the mouse had only liver tumors, the F-344 rat had a neoplasm of some type in 16 out of the 21 bioassays and liver tumors in 8 of those bioassays. When the mouse had both liver and nonhepatic tumors, the F-344 rat had a neoplasm of some type in 8 out of 10 bioassays and liver tumors in 4 of those bioassays. The mouse had tumors of some type in 12 bioassays, while the F-344 rat had no tumors. The F-344 rat had tumors of some type in 19 bioassays, when the mouse had no tumors.

A comparison of tumor occurrence in the B6C3F1 mouse and the OM rat when used in the same bioassay is presented in Table 6. The mice had only liver tumors in 16 bioassays. In those bioassays the OM rat had tumors at some site in 2 of the 16 bioassays with liver tumors found in both bioassays. The mouse had liver and nonhepatic tumors in 3 bioassays and the OM rat had only nonhepatic tumors in 2 of those bioassays. The mouse had some type of tumor in 16 bioassays in which the OM rat had no tumors. The OM rat had some type of tumor in one bioassay when the mouse had no tumors.

There were liver neoplasms in the B6C3F1 mouse in 50 of the 85 bioassays where at least one sex of one species had some type of tumor. They occurred in both sexes in 25 bioassays, only in the male in 7 and only in the female in 18.

In summary, the following points should be emphasized.

1. The B6C3F1 mouse had a high incidence of liver tumors in control animals, especially in the male. This incidence

TABLE 5. Tumors[a] in F-344 Rats or B6C3F1 Mice

|  |  | F-344 rat: | | | |
|---|---|---|---|---|---|
| B6C3F1 Mice | | Liver tumors only | Liver & nonhepatic tumors | Nonhepatic tumors only | No tumor |
| Tumor type | Number | | | | |
| Liver tumors only | 21 | 1 | 7 | 8 | 5 |
| Liver and nonhepatic tumors | 10 | 3 | 1 | 4 | 2 |
| Nonhepatic tumors only | 10 | 0 | 0 | 5 | 5 |
| No tumors | 19 | 4 | 0 | 15 | 0 |
| Total | 60 | 8 | 8 | 32 | 12 |

[a]Significantly increased tumors in at least one sex.

TABLE 6. Tumors[a] in OM Rats or B6C3F1 Mice

|  |  | OM rat: | | | |
|---|---|---|---|---|---|
| B6C3F1 mice | | Liver tumors only | Liver & nonhepatic tumors | Nonhepatic tumors only | No tumor |
| Tumor type | Number | | | | |
| Liver tumors only | 16 | 1 | 1 | 0 | 14 |
| Liver and nonhepatic tumors | 3 | 0 | 0 | 2 | 1 |
| Nonhepatic tumors only | 5 | 0 | 1 | 3 | 1 |
| No tumors | 1 | 0 | 0 | 1 | 0 |
| Total | 25 | 1 | 2 | 6 | 16 |

[a]Significantly increased tumors in at least one sex.

varied significantly between control groups in different laboratories.
2. When testing chemicals at the maximum tolerated dose, approximately one half of the chemicals (94 of 192) did not cause an increase in tumors at any site in either sex of two rodent species.
3. Of the 85 chemicals carcinogenic in at least one sex of one species, 59% (50 of 85) caused liver tumors in the B6C3F1 mouse, 27% (16 of 60) caused liver tumors in F-344 rats and 12% (3 of 25) caused liver tumors in OM rats. Twenty of 85 chemicals caused only liver tumors in the B6C3F1 mouse and caused no tumors in either F-344 or OM rats.
4. When the B6C3F1 mouse had liver tumors, the F-344 rat had tumors at some site in 24 of 31 bioassays and the OM rat had tumors at some site in 4 of 19 bioassays.

In conclusion, the importance of the mouse liver tumor in the interpretation of these bioassays has been shown. Research on the mechanisms of liver tumor induction in the B6C3F1 mouse is urgently needed so the significance of these bioassay results for humans can be determined.

REFERENCES

1. Chu KC, Cueto C Jr., Ward JM: Factors in the evaluation of 200 National Cancer Institute bioassays. J Toxicol Environ Health 8: 251-280, 1981.

2. Sontag JM, Page NP, Saffioti V: Guidelines for carcinogen bioassay in small rodents. NCI-CG-TR-1 USDHEW, Stock Number 017-042-00118-8, Superintendent of Documents, U. S. Government Printing Office, Washington, D.C.

3. Griesemer RA, Cueto C: Toward a classification scheme for degrees of experimental evidence for the carcinogenicity of chemicals for animals. In: Molecular and Cellular Aspects of Carcinogen Screening Tests, edited by R Montesano, H Bartsch, L Tomatis. Lyon: IARC Scientific Publication no. 27, 1980.

4. Ward JM, Goodman DG, Griesemer RA, Hardisty JF, Schueler RC, Squire RA, Strandberg JD: Quality assurance for pathology in rodent carcinogenesis tests. J Environ Pathol Toxicol 2: 371-378, 1978.

5. Tarone RE, Chu KC, Ward JM: Variability in the rates of some commonly occurring tumors in Fischer-344 rats and (C57BL/6N x C3H/HeN)F1(B6C3F1) mice. J Natl Cancer Inst 66: 1175-1181, 1981.

6. Ward JM, Goodman DG, Squire RA, Chu KC, Linhart MS: Neoplastic and non-neoplastic lesions in aging C57BL/6N x C3H/HeN/F1 (B6C3F1) mice. J Natl Cancer Inst 63 849-853, 1979.

7. Goodman DG, Ward JM, Squire RA, Chu KC, Linhart, MS: Neoplastic and non-neoplastic lesions in aging F-344 rats. Toxicol Appl Pharmacol 48: 237-248, 1979.

8. Goodman DG, Ward JM, Squire RA, Paxton MB, Reichardt WD, Chu KC, Linhart MS: Neoplastic and non-neoplastic lesions in aging Osborne-Mendel rats. Toxicol Appl Pharmacol 55:433-447, 1980.

Chapter 3

# Implications of Mouse Liver Neoplasms for Predicting Carcinogenicity

**Robert A. Squire**

There exists considerable disagreement about the induction of mouse liver neoplasms as predictors of human risk. This is part of the reason for bringing these chapters together. Since concensus on this issue is lacking, it is appropriate to examine the type of information that is available and necessary for predicting carcinogenic risk in humans. Since the enactment of the Delaney clause and particularly in the last decade, there has been considerable progress in the field of experimental carcinogenesis. Scientific evidence and opinion have modified somewhat since the regulatory hearings on DDT, aldrin and dieldrin, in which controversy over the significance of mouse liver tumor induction surfaced as a major issue. Perhaps the most important observation that has emerged is that there is a great diversity among chemical carcinogens. This diversity involves several properties, including:

1. the chemical structure and biological reactivity of animal carcinogens;
2. the ability of some chemicals to interact with and damage genetic material in cells;
3. the widely varying potency of chemical carcinogens in animals; and
4. the prevalence of the effects, i.e., the number of different species or tissues that are susceptible.

Although proposed mechanisms of neoplastic transformation remain hypothetical, there is evidence that some chemicals or their metabolites may induce cancer by direct mutagenic-like effects in cells. Others apparently act in a secondary or indirect manner, which is poorly understood. These indirect-acting agents often require high, toxic exposures to produce a carcinogenic effect. We have also learned that there is great variation in the metabolic and pharmacokinetic pathways by which various animal species handle chemicals, and that detoxification and repair mechanisms exist that may preclude the carcinogenic effect of chemicals. Thus, the induction of tumors in test animals may have many _different_ implications, and any specific test result, such as mouse liver tumor induction, should be considered as part of the total toxicological evidence rather than in isolation.

The attention and controversy surrounding this particular toxicological endpoint persist because of several questions that, in the view of some, remain unanswered. Three of the most fundamental questions are:

1. Are the hepatic lesions in mice that are under discussion actually neoplasms?
2. Is the induction of these hepatic lesions in mice qualitatively relevant to potential human cancer risk?
3. Is the mouse an appropriate surrogate for quantitative risk assessment?

I think it can be safely stated that the answers to these questions remain controversial. The primary obstacle encountered in addressing the first question is the difinition of neoplasia itself. There are numerous definitions of cancer or neoplasms in different textbooks and articles and in some well ublicized committee reports, but no two are identical. The problem stems from the fact that the phenomenon of neoplasia is not understood. There are no absolute markers for neoplastic cells, and the identification of a neoplasm, at least in its early stages, depends upon the experience and judgement of the clinician or pathologist. The basis for a pathologist's judgement is correlation of morphological features with biological behavior, since biological behavior is the ultimate criterion for establishing the neoplastic nature of the lesion. This correlation is often, of necessity, based upon predicted rather than observed behavior, since the pathologist is called upon to classify lesions before they express their full biological potential. To be absolutely certain that a tissue alteration is cancer, one would have to wait until invasion or metastasis developed and was discovered, since these are the biological features that characterize malignancy. This is hardly practical under most circumstances. Furthermore, some neoplasms are lethal due to their location or functional impairment before they ever invade or metastasize.

In animal testing, there has been pressure to push the frontiers of tumor recognition back further and further to the identification of early neoplastic or preneoplastic lesions and now even to "putative preneoplastic" cellular alteration (1). In making these judgments, it is important to realize that we are often dealing with areas of considerable uncertainty.

There are lesions in the mouse liver that virtually all pathologists will agree upon, and these are the fairly typical hepatocellular carcinomas (2). They look like carcinomas in any other species. Even without demonstrated invasion or metastasis, the lesions often show sufficiently atypical features to warrant a consensus diagnosis. The lesions that are primarily in question are the hepatocellular foci and nodules (2). I think it important to recognize that these lesions in mice are similar morphologically to lesions in rats, humans, and other species. That is to say, the mouse is not unique since there exist pathological uncertainties in other species and even in other tissues concerning similar lesions. The hepatocellular nodules, also called Type A nodules, hepatocellular adenomas, hyperplastic nodules, neoplastic nodules, etc., are analagous to nodular lesions in many animals. They are proliferative lesions that often precede the ultimate development of hepatocellular carcinoma (3-5). They may persist after the provoking stimulus is removed. They are biochemically and morphologically altered (6) and transplantable (7). A recent report provides evidence that some rat liver nodules have the capacity for invasion of normal tissue in vitro, providing further evidence of their neoplastic nature (8).

The hepatocellular foci are a different question, however. They have been characterized in detail, mainly in rats, and really never received major attention until recently. It is evident that they are also altered cells, that is, altered morphologically, biochemically, and enzymatically (6). However, all cellular markers to date are empirical. They reflect, but they do not necessarily define, the fundamental nature of the alteration. It is also reocgnized that foci occur in normal control rodents. There are no studies that conclusively show that foci can progress to neoplasms or that they are transplantable. If one wishes to use the term "preneoplastic," then it should be made clear that this implies a chronological but not necessarily an intrinscially related precursor, until further evidence is accumulated. In summary, suffice it to say that cntroversy still exists and that the pathologic classification of some mouse liver lesions has yet to be agreed on.

The second question, "Is the induction of these hepatic lesions in mice relevant to potential human cancer risk?", is of equal, or perhaps greater, importance. One concern has been that this and the first question may be confused. Those who do not believe that the mouse is an appropriate surrogate for human risk assessment may, in certain instances, discount the biological implications of the liver lesions to the mouse itself in order to make their point. The two questions should be kept distinct and separate. If the induction of liver cancer in the mouse is not considered relevant to human risk, then that issue should be argued, rather than the pathologic interpretation and its significance to the mouse. Perhaps, in fact, part of the controversy stems from the different perspectives brought to the issue - those of veterinary pathologists versus medical pathologist and their differing patient orientations.

In answering the question of the relevance of mouse liver neoplasms, it is useful to examine the basis for the use of test animal surrogates in general. The rationale for the use of animals in toxicological testing is similar to the rationale for the use of animals in all biomedical research and may be summarized as follows:

1. Mammals are anatomically, physiologically, and biochemically similar.

2. Mammals have similar health and disease manifestations and causes.
3. Mammals respond similarly to exogenous chemical, biological, and physical agents. Differences are primarily quantitative rather than qualitative.

There are exceptions to these statements throughout the animal kingdom; however, they are the general rule. The question may therefore be asked, "Is the induction of mouse hepatic neoplasms predictive of cancer risk in any other species - not only human?" One may examine the correlation of mouse liver tumor induction with the induction of tumors at other sites and in other species. This has been done on two occasions, and the most recent survey by Ward et al. (9) was based partly upon the results of chemical tests from the NCI Carcinogenesis Testing Program (also see Chapter 2). Of 53 chemicals from the NCI program, all of which were hepatocarcinogenic in mice, only 55% were also carcinogenic either in other tissues of mice or in rats. Thus, the correlation was not sufficiently high to resolve the issue. As the authors state, "The interpretation and significance of these findings awaits further research into the mechanisms of carcinogenesis." It is probably safe to say that if or when we understand the mechanisms of carcinogenesis, many of the problems we are dealing with at this symposium will no longer exist.

The question is not whether the classical, potent animal carcinogens may induce liver cancer, because one would expect that they might. The question is rather, "Are there chemicals that would not be carcinogenic even at high exposure levels except in mouse liver?" Ward et al. (9) also correlated their findings with mutagenesis testing results in bacteria, and again the correlations were partial. Some of the chemicals that caused liver tumors in mice, but not in other tissues or species, were mutagenic and some were not. It may well be that such correlative studies will not resolve the questions surrounding mouse liver neoplasms, or any other specific animal neoplasms that are singled out to examine for relevance to human risk.

Most impressive is the spontaneous incidence of liver neoplasms in many inbred strains of mice (10). Even in those strains in which the incidence is reportedly rate (that is, less than 1%) it is still far higher than encountered in the human population. A 1% cancer rate would be a major epidemic in humans! We can thus conclude that, according to the current knowledge of carcinogenic mechanisms, there must be a high population of latent neoplastic or "initiated" liver cells in mice, and this apparently is not the case in humans, at least in this country. This may indicate a qualitative difference between the mouse and the human in their susceptibilities to liver cancer, particularly as the result of exposure to modifying or "promoting" factors. Experimental induction of tumors that have such high natural occurrence in test animals is probably less relevant to human risk than is the induction of tumors that are normally rare in the test animals. Obviously, mechanisms must be examined and would have to include the possible roles of genetic susceptibility and/or oncogenic retroviruses, which are so prevalent in mice. If specific murine viruses are responsible for the initiation of liver tumors, as they may be for lung tumors and are for mammary tumors and leukemias in mice (10), then one may question whether the mouse is an appropriate surrogate, even qualitatively, for determining carcinogenic risk from exposure to environmental chemicals. Of course, similar roles for human viruses would also have to be considered, and it has already been established that there is an apparent association between hepatitis B virus and human liver cancer (11).

At present we do not know what role, if any, viruses are playing in the natural incidences and/or susceptibility of mice to hepatocarcinogenesis. However, it has been established that oncogenic viruses are ubiquitous in mouse tissues. Further research efforts should be devoted to investigating viral chemical interactions and cocarcinogenic effects. Also, the extensive knowledge of mouse genetics could be applied to studies of cancer susceptibility.

The final question relates to the appropriateness of the mouse for risk assessment. This is ultimately an important part of regulatory decisions, whether through formal application of mathematical models, the use of safety factors, or, implicity, through informed judgment. Here again, a very important consideration is the background incidence of liver tumors in inbred mice. A recent report by Tarone et al. (2) shows the incidence and variation of liver tumors among five different test laboratories within the NCI Testing Program. It can be seen that there is great variation, with the incidence ranging from 0 to as high as 58% in control male mice. These were contemporary historical controls, which points out the hazards of false negatives and false positive findings if total reliance is placed upon biostatistical analysis rather than biological judgment. Although statistical evaluation takes into account matched control incidence, it does not take into account the implications of carcinogenic mechanisms and susceptibility when extrapolating the results to human risks. The factors involved in the induction of liver tumors in mice as the result of exposure to chemicals are probably not operating in the same way in humans, since in humans the liver cancer background rate is comparatively low. Whether we speak in terms of "initiation" and "promotion," or concede that the mechanisms of carcinogenesis, although uncertain, are complex and probably multiple, those mechanisms at work in the mouse are apparently not at work in humans to the same degree.

If the mouse liver tumor is presently accepted as a valid toxicological endpoint, which I believe it must be, it may still not be appropriate to apply such findings to quantitative risk assessment, for the reasons already stated. Of course, one could make the same argument for several different tumor endpoints in laboratory rodents, since most of the natural incidences exceed those in humans at virtually all tissue sites. However, mouse liver tumors, like leukemia, mammary tumors, and lung tumors in certain strains, are extreme examples and certainly do not quantitatively reflect the human risk situation.

Human risk assessment from experimental carcinogenesis data must include the total weight of evidence and must not be limited to any single animal finding. I recently proposed a semiquantitative ranking system as an example of how this could be accomplished (13). First to be considered are the nature and extent of the animal findings. For example, how many animal species and tissue sites are affected? If several of these are positive, it is more likely that humans will be similar in response rather than uniquely different. If, on the other hand, the mouse liver is the only target, even after multiple-species testing, perhaps that chemical should be regarded differently and with less concern. Most of the established human carcinogens that have been adequately tested in animals are carcinogenic in multiple species and at multiple sites (14). As already indicated, we must also consider the spontaneous incidences in control animals of the neoplasms in question, since this reflects the relative sensitivity of the test animal to the induction of such neoplasms. In the case of mouse liver tumors, we may be dealing with over-sensitive indicators of risk, and some formalized procedure for taking this into consideration in quantitative risk assessment may be necessary.

The nature of test chemical, including its biological reactivity and genotoxicity, must also be considered. If it is demonstrated that a chemical or its metabolite would not interact in a direct way with cellular DNA, then an indirect mechanism must be postulated and may require a certain level of toxic alteration before neoplastic transformation would occur. The role of cellular proliferation in carcinogenesis is well demonstrated (15), and a tissue such as the mouse liver, which apparently contains a large population of so-called "initiated" or latent neoplastic cells, is probably more susceptible to the nonspecific, proliferative effects of liver damage than a tissue in which the background rate of cancer is very low.

All this is not to say we can dismiss mouse liver neoplasms as significant endpoints in risk assessment. There are reasons, however, to question whether mouse liver tumors are as appropriate as are some other animal results for regulatory purposes, particularly for quantitative risk assessment. Although much of the burden of proof remains, tools are available to answer at least some of the questions. It is very possible that specific, targeted research would be more efficient and productive, and even less costly, than the continuing debate over mouse liver tumors, which seems unable to resolve the issues with the information at hand.

REFERENCES

1. Ogawa K, Solt DB, Farber E: Phenotypic diversity as an early property of putative preneoplastic hepatocyte pouplations in liver carcinogenesis. Cancer Res 40:725-733, 1980.

2. Frith CH, Ward JM: A morphologic classification of proliferative and neoplastic hepatic lesions in mice. J Environ Path Toxicol 3:329-351, 1980.

3. Ward JM, Blahakis G: Evaluation of hepatocellular neoplasms in mice. J Natl Cancer Inst 61:807-811, 1978.

4. Lipsky MM, Hinton DE, Klaunig JE, Trump BF: Biology of hepatocellular neoplasia in the mouse. I. Histogenesis of safrole-induced hepatocellular carcinoma. J Natl Cancer Inst 67:365-376, 1981.

5. Frith CH, Codell RL, Littlefield NA: Biologic and morphologic characteristics of hepatocellular lesions in Balb/c female mice fed 2-acetylaminofluorene. J Environ Pathol Toxicol 3:121-138, 1979.

6. Lipsky MM, Hinton DE, Klaunig JE, Goldblatt PJ, Trump BF: Biology of hepatocellular neoplasia in the mouse. II. Sequential enzyme histochemical analysis of Balb/c mouse liver during safrole-induced carcinogenesis. J Natl Cancer Inst 67:377-392, 1981.

7. Williams GM, Hirota N, Rice JM: The resistance of spontaneous mouse hepatocellular neoplasms to iron accumulation during rapid iron loading by parenteral administration and their transplantability. Am J Pathol 94:65-74, 1979.

8. Wanson JC, Ridder LD, Mosselmans R: Invasiveness of hyperplastic nodule cells from diethylnitrosamine-treated liver. Cancer Res 41:5162-5175, 1981.

9. Ward JM, Griesemer RA, Weisburger EK: The mouse liver tumor as an endpoint in carcinogenesis tests. Toxicol Appl Pharmacol 51:389-397, 1979.

10. Squire RA, Goodman DG, Valerio MG, Frederickson T, Strandberg JD, Levitt MH, Lingeman CH, Harshbarger JC, Dawe CJ: Tumors. In: Pathology of Laboratory Animals, edited by K Benirschke, FM Gardner, TC Jones. New York: Springer-Verlag, 1978.

11. National Institutes of Health international workshop on hepatitis B and liver cancer. Cancer Res 37:4672, 1977.

12. Tarone RE, Chu KC, Ward JM: Variability in the rates of some common naturally occuring tumors in Fischer 344 rats and (C57BL/6N X C3H/HEN) F1 (B6C3F1) mice. J Natl Cancer Inst 66:1175-1181, 1981.

13. Squire RA: Ranking animal carcinogens: A proposed regulatory approach. Science 214:877-880, 1981.

14. Tomatis L, Agthe C, Bartsche H, Huff J, Montesano R, Saracci R, Walker E, Wilborn J: Evaluation of the carcinogenicity of chemicals: A review of the monograph program of the international agency for research on cancer (1971-1977). Cancer Res 38:877-885, 1978.

15. Schulte-Hermann R, Ohde G, Schuppler J, Timmermann-Trosiener I: Enhanced proliferation of putative preneoplastic cells in rat liver following treatment with the tumor promoters phenobarbital, hexachlorocyclohexane, steroid compounds, and nafenopin. Cancer Res 41:2556-2562, 1981.

Chapter 4

# Genetic, Hormonal, and Dietary Factors in Determining the Incidence of Hepatic Neoplasia in the Mouse

**Paul Grasso**

INTRODUCTION

The induction of an increased incidence of tumors is universally regarded as a reliable index of carcinogenic activity. Indeed, despite the substantial and often spectacular advances in genetic toxicology in recent years, the only means available to assess the carcinogenic activity of chemicals is to find out whether they increase the incidence of tumors in appropriate studies on experimental animals.

The mouse has been a favorite in animal experimental studies for a number of reasons. Its small size, prolific rate of reproduction, assumed sensitivity to carcinogens, and ready availability at a reasonable cost have considerably assisted in perpetuating the mouse as the species of choice for conducting carcinogenicity studies.

Over the last three or four decades, a considerable number of chemicals have been tested in the mouse. Many of these appeared to induce an increase in the incidence of hepatic tumors in the treated mice compared with controls. In reviews published by Tomatis et al. (1) and Ward et al. (2), the view was expressed that induction of mouse hepatic neoplasia is a good and reliable index of carcinogenic activity, as valid as the evidence produced in the rat and hamster at any site.

I think that this conclusion is understandable in light of the experience then available. However, since then evidence has accumulated that a number of factors may cause an unpredictable variation in the background incidence of mouse hepatic tumors, which could seriously interfere with the interpretation of results. However, there was already in existence a substantial amount of evidence that such factors might exist. Because of a revival of interest in the question of the validity of mouse hepatic tumors as an index of carcinogenicity, further experimental evidence has been generated and has placed the opinions and conclusions of the earlier workers on a better scientific basis.

For this presentation, we review a selection of the available data on three factors thought to exercise an important influence on mouse hepatic neoplasia and to perhaps determine its incidence in any particular experiment. No distinction is made in most of the early literature between nodular hyperplasia, adenoma, and carcinoma. In order to present some sort of coherent picture it was necessary to consider all hepatic nodular lesions as neoplastic, even though I am convinced that this assumption is unlikely.

EFFECTS OF DIET

Let us first consider the effect of diet on mouse hepatic tumors. The first observations were made by Tannenbaum (3) and resulted from an interest in the role that nutrition played in determining cancer incidence. He must have observed the mice developing an unsightly obesity as well as a variety of tumors as they grew older and wondered whether they would develop the same type of tumors if they were kept "lean and fit."

In an early experiment, Tannenbaum (3,4) reduced the dietary intake of C3H male mice from 4.0 gm/day to 3.2 gm/day. This caused the hepatoma incidence to drop from approximately 55% to approximately 15%. This observation clearly indicated the profound influence that dietary restriction plays on the incidence of spontaneous tumors.

This topic has been investigated in greater depth recently. Tucker (5) showed that by reducing the dietary intake of mice by about 20%, they lived longer and had fewer hepatic tumors than did the mice fed _ad libitum_. As shown in Table 1, the effect was more marked in the liver of males than of females. In fact, the incidence in the female was not affected. This data suggests a "basal" incidence that is unaffected by diet, with the incidence in the male brought down to "basal" level by the reduced diet. However, this does not mean that the female incidence of hepatoma cannot be altered by manipulation of the diet.

In this investigation, not only the hepatic tumors, but tumors of other tissues had a lower incidence in the food-restricted mice (5). This reduction included tumors of the lung, lymphoid system, and pituitary; tumors which occur commonly in mice.

A reduction in the incidence of the same tumors was observed by Conybeare (6). As shown in Table 2, both lung and lymphoid tumors were reduced when diet was restricted by 25%, but the liver tumors showed the most dramatic drop.

These experiments leave little doubt that dietary restriction plays an important part in reducing the development of mouse hepatic neoplasia. It is pertinent to inquire whether any particular

TABLE 1. Dietary Restriction and Tumor Incidence in Charles River CD Mice

|  | Total[c] | Liver | Pit | Lung | Lymph |
|---|---|---|---|---|---|
| Males |  |  |  |  |  |
| Ad libitum | 40 | 14 | 1 | 8 | 10 |
| Restricted | 26 | 2 | 0 | 3 | 5 |
| Females |  |  |  |  |  |
| Ad libitum | 41 | 2 | 7 | 7 | 17 |
| Restricted | 23 | 4 | 0 | 3 | 10 |

[a] Data from Tucker (5).
[b] Groups of 50.
[c] Animals with tumors.

component of the diet may be implicated in influencing the incidence of hepatic tumors.

There is evidence that some proteins and vegetable oils may increase the incidence of liver tumors when given in large amounts in the diet. Whether these play a specific role in the reduction of the tumor incidence in dietary restriction experiments is open to question.

The role of proteins in determining tumor incidence was first investigated by Tannenbaum and Silverstone (7,8). Five groups of

TABLE 2. Reduction of Tumor Incidence in Mice by Simple Dietary Restriction[a,b]

|  | Lung | Liver | Lymphoma | Others |
|---|---|---|---|---|
| Males |  |  |  |  |
| Ad libitum | 30 | 47 | 4 | 8 |
| Restricted | 19 | 12 | 1 | 4 |
| Females |  |  |  |  |
| Ad libitum | 24 | 7 | 11 | 12 |
| Restricted | 8 | 1 | 4 | 4 |

[a] Data from Conybeare (6).
[b] Restricted = 75% of ad libitum consumption. Ad libitum consumption = 5.0 gm/day/mouse.

13-14-week-old C3H female mice were fed diets containing 9, 18, 27, 36, or 45% casein, a complementary amount of corn starch, and identical amounts of other constituents, which included essential amino acids, fats, vitamins, and mineral salts. The hepatic tumor incidence was 10, 38, 43, 50, and 33% in the respective groups. An increase in the incidence of hepatic tumors was also found in male C3H mice. Only three dietary groups, containing 9, 18, and 45% casein appeared to have been investigated. The hepatic tumor incidence was 11, 61, and 38%, respectively.

The lower incidence of tumors with the highest percentage of casein (45%) in the diet cannot be readily explained. However, the body weight of this group of animals was the same as that in the 9% casein group, while that of the other groups was substantially higher. It would seem likely that the amount of casein in the group given 45% casein may have made the diet unpalatable to rats, but we have no data on food consumption.

This experiment shows that some proteins increase the hepatic tumor incidence without the intervention of any other chemicals. Other proteins may have a different effect. For example, a diet containing 9% casein and 9% gelatin did not yield the same tumor incidence as 18% casein. On the other hand, supplementing a 9% casein diet with small amounts of methionine and cystein augmented the hepatoma incidence to that produced by 18% casein.

Apart from proteins, the other dietary constituent that has been clearly shown to influence the incidence of mouse hepatic neoplasia is the fatty content of the food. Thus Tannenbaum and Silverstone (7) showed that a tenfold increase in the fat content of the diet to a group of C3H mice resulted in an increase in hepatoma incidence from 36 to 51%. The fat used in this experiment has been described as partially hydrogenated cottonseed oil and care had been taken to ensure that the caloric intake of the two groups of animals was the same (8).

Somewhat similar results were obtained by Gellatly in 1975 (9) in C57BL mice (Table 3). These results are of interest because this strain of mice has a much lower incidence of naturally occurring hepatic tumors than the C3H mice employed by Tannenbaum and Silverstone (7). They show that dietary influence on tumor production is present in strains with a low, as well as in strains with a high, incidence of hepatic tumors.

One important conclusion can be drawn from these experiments. The caloric content of the diet, as well as the amount of fat and protein, may influence the incidence of hepatic tumors in mice. It is probable, although not certain, that these effects may occur in mice with either a high or low natural incidence of liver tumors. It is likely that the puzzling increase in the incidence of hepatoma in singly caged mice can be explained on the basis of excessive food consumption. Mice caged singly have less interference with feeding than mice caged in groups (10).

TABLE 3. Influence on Groundnut Oil on Hepatoma Incidence in C57BL Mice[a]

| Diet | Fat Content | Number tested Male | Number tested Female | % Incidence Male | % Incidence Female |
|---|---|---|---|---|---|
| Stock pelleted | 2% | 276 | 260 | 10 | 12 |
| Purified | 5% | 64 | 74 | 30 | 30 |
| Purified | 10% | 101 | 121 | 39 | 64 |

[a] Data from Gellatly (9).

GENETIC FACTORS

The genetic control of hepatic neoplasia is not as thoroughly understood as that of lymphoma and pulmonary adenoma in the mouse. There are a number of points, however, that tend to indicate that genetic factors may influence the evolution of mouse liver tumors.

The first indication of a genetic effect is the very large difference in hepatic tumor incidence that exists between the various strains of mice (11). For example, C3H mice have an incidence which may be as high as 40-60%, while C57BL mice may have an incidence as low as 4-5% (12). This variation in incidence of neoplasms may be noted for other tumors. Spontaneous pulmonary adenomas are relatively common in strain RF mice (approximately 50%), while they occur in a relatively low incidence in other strains, for example C57BL (12). It is now generally accepted that the development of pulmonary adenoma in the mouse is controlled by at least one pair of genes and that the expression of these genes accounts for the large interstrain differences in its appearance (13). The large interstrain differences in the incidence of hepatic neoplasia could be explained on a similar basis.

There are other factors which point to gene control in the development of spontaneous hepatic tumor. It has been known for decades that the incidence of hepatic tumor changes over several generations. Andervont (14) first showed that, within 2-3 year, a noticeable change occurred in the tumor incidence of the C3H and CBA mice (Table 4). This was also the experience of Tucker at ICI. As shown in Fig. 1, an increase in the incidence of tumors in the ICI strain of mice was noted over a 7-year period of observation (cited in Grasso and Hardy (11)).

TABLE 4. Examples of Drift in the Incidence of Hepatoma in Mice[a]

| Strain | Year | Sex | Number tested | Tumors (%) |
|---|---|---|---|---|
| C3H | 1942 | F | 362 | 11 |
|  | 1943/1945 | F | 179 | 5 |
|  | 1943 | M | 43 | 12 |
|  | 1944 | M | 86 | 55 |
| CBA | 1947 and 1978 | M | 68 | 20 |
|  |  | M | 40 | 3 |
|  |  | F | 63 | 5 |
|  |  | F | 29 | 3 |

[a] Data from Andervont (14).

This slow change over the years recalls the so-called "genetic drift," referred to by geneticists, which is widely acknowledged to occur in inbred strains (15). In theory, the genetic

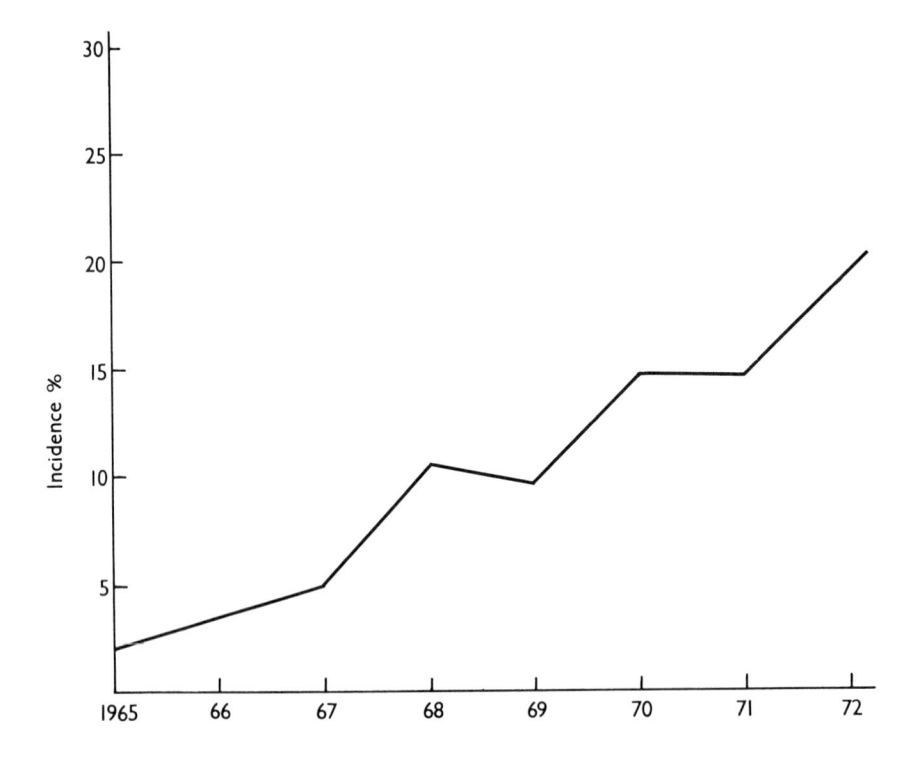

FIGURE 1. Hepatic tumors in ICI mice (untreated).

GENETIC, HORMONAL, AND DIETARY FACTORS 53

characteristics of an inbred line of mice is stable, so that the heritable characteristics remain unaltered from one generation to the next. In practice, it is observed that some alterations in the genetic traits do occur, including the genes that control tumor incidence. Genetic drift is seemingly unavoidable because, despite the precautions taken, it is impossible to completely control the transmissible genetic traits. In my view, if this drift occurs when precautions are taken to prevent it, there is all the more reason to expect it to occur when precautions are absent. This would account for the marked change in the incidence of hepatomas observed by Andervont and Tucker.

The other evidence is from Heston's experiments (16), which indicate that the so-called viable yellow gene is associated with a high (100%) incidence of hepatoma.

It would seem that there are indications of a genetic influence on the development of mouse liver tumors. I would not like to put it any more strongly, because other factors might affect incidence over the years.

HORMONAL FACTORS

The third factor that might be involved in the process of development of the hepatic nodular lesion or "hepatoma" is hormonal. The discussion of this factor in relation to hepatoma is not surprising, since the topic of hormones causing cancer is well established in organs that are under hormonal regulation and has been the subject of intensive study for several years. This is particularly so in the case of the mammary gland, but there is ample evidence that tumor formation in other organs, such as ovary, uterus, and prostate, is affected (17).

Strictly speaking, the liver is not an endocrine organ. However, it is known that cortisol is metabolized in the liver and is probably an important site for the metabolism of other hormones, in view of the special function that the liver possesses in the biotransformation of most endogenous compounds. Furthermore, there is evidence that glucocorticoids may play an important role in controlling the proliferative activity of the liver. Thus, in an experiment carried out by Desser-Weist (19), it was shown that reduction of the corticosterone level by adrenalectomy increased the rate of thymidine incorporation into the DNA, the mitotic index and the percentage of tetraploid cells.

There is also evidence that progesterone given by injection stimulates mitotic activity in the liver. It is well known that in the pregnant rat, where there is a significant increase of progesterone, the weight of the liver, the rate of mitosis, and the DNA, RNA, and protein content are increased (19,20).

Despite these interesting findings, there is no clear evidence of the role these hormones play in the production of mouse hepatic neoplasia. Certainly they attracted the attention of the early workers, presumably because of the well-authenticated observation that the incidence of liver tumors is higher in the male than female mouse, irrespective of the strain (11).

One of the earliest observations was that of Burns and Schenken (21,22) in 1943. They compared the hepatoma incidence in breeders and virgins and reported an incidence of 27% in 60 breeding C3H males, in contrast to an incidence of 6% in 16 virgin males. In the females, the incidence was 0% in 47 breeders, in contrast to an incidence of 10% in 10 virgins. They commented that "breeding unfavorably affects the development of tumors in the males but favorably influenced the development of tumors in the female."

Other comparisons have been made by different authors on the effect of breeding on the hepatoma incidence in different strains of mice. These results have been summarized by Murphy (23) and they indicate that pregnancy does not appear to have any discernible influence on the incidence of hepatomas in female mice, since virgin and breeder mice seem to be at the same risk of developing this tumor. The results are inconclusive, but the general picture tends to indicate that pregnancy has a beneficial effect of lowering the incidence of hepatomas. This provides some tentative evidence that perhaps progesterone is not as important in favoring the development of hepatomas as one might think from an appraisal of the short-term hepatic effects it produces.

Some systematic studies have been conducted on the possible role of the sex hormones in the induction of hepatic neoplasia in the mouse. Miller and Pybus (24) gave weekly subcutaneous injections of 0.05 ml of a 0.06% solution of estrone in olive oil to gonadectomized and intact CBA mice of both sexes. Under this regimen, there was an increased incidence of hepatomas in intact males but a decreased incidence in gonadectomized males. However in females, the tumor incidence was reduced in both intact and ovariectomized animals. Unfortunately, the death rate among treated animals was greater than in controls, so the validity of this interesting observation is uncertain.

Schenken and Burns (25) administered testosterone propionate to male and female mice, but the mortality rate among the treated mice was too high to allow any conclusions to be drawn. These authors also implanted cholesterol pellets containing either diethylstilbestrol or estradiol in male mice, resulting in a reduced hepatoma incidence. Unfortunately, this interesting result was noted in only a small number of mice and was not investigated further.

Andervont (14) found that castration reduced, by a third, the incidence of hepatomas in C3H mice but had little influence on the hepatoma incidence in CBA mice. Agnew and Gardner (26) reported that the incidence of hepatic tumors was increased in female C3H mice treated with testosterone.

The hormonal factors that may have an effect on hepatic neoplasia in the mouse may be summarized as follows:

1. Castration reduced the incidence in males.
2. Estrone reduced the incidence in gonadectomized males and females but increased it in intact males.
3. Pregnancy (progestrone) reduced the incidence in females.
4. Testosterone increased the tumor incidence.
5. Estradiol and DES decreased the incidence in intact male mice.

Other hormones are known to have an effect on the liver, particularly thyroxine and growth hormone. When administered to intact animals, both cause an increase in liver size and ploidy (27). The effect they have on the induction of liver tumors is unknown, but we can make two points. First, a number of insecticides that break down to thiourea inhibit thyroid hormone function but also induce liver tumors in mice (18,28). This would suggest that reduction of thyroid hormone by itself is not likely to reduce the risk of tumor induction. Second, there is no experimental evidence to suggest the way in which growth hormone might influence hepatic neoplasia in the mouse. There is, however, an interesting observation by Tucker (5) and by Conybeare (6) that reduced caloric intake in the mouse resulted in a dramatic reduction of both pituitary and hepatic tumors. It would be interesting to investigate whether the reduction in hepatic tumors is in some way connected with the reduction in pituitary adenomas via a reduction in growth hormone (33). However, one must recognize that the pituitary reduction occurs in females, while hepatomas are reduced in the males.

DISCUSSION AND SUMMARY

The factors that influence tumor production can be summarized as follows: (1) Caloric restriction reduces the incidence of naturally occurring hepatic tumors. (2) Estrone also seems to reduce the incidence in females and in gonadectomized males, but paradoxically it seems to increase the incidence in intact males. (3) A high protein or fat content of the diet leads to an increased incidence of hepatic tumors. (4) Genetic influence is uncertain, but there are indications that some genetic factors are involved in tumor formation.

The question that might be raised at this stage is whether dietary, hormonal, and genetic factors have any relevance to the assessment of toxicological data or to choosing between the mouse and other species in carcinogenicity testing. In addition, the influence of dietary, hormonal, and genetic factors on the induction of hepatic tumors by carcinogens should be considered.

The evidence regarding the effect of caloric restrictions on the induction of hepatic tumors is conflicting. In an early experiment, Kirby (29) demonstrated that dietary restriction of about 25% almost totally abolished the carcinogenicity of aminoazotoluene. More recent experiments by Tucker (personal communication in 1980) showed that dietary restriction altered the hepatocarcinogenic response to ICI. 55,695 but did not abolish it altogether (Table 5). Obviously, this field urgently requires further investigation.

TABLE 5. Hepatic Response to the Carcinogen ICI 55,695 in Male Mice[a,b]

| Response | Ad libitum feed | Restricted feed |
|---|---|---|
| Hyperplastic nodules | 7 | 15 |
| Benign | 14 | 10 |
| Malignant | 10 | 1 |

[a] Data from Tucker (5).
[b] Groups of 50.

The effect of hormonal and genetic factors on chemically-induced mouse liver tumors is far from clear. However, two experiments may throw some light on the effect of sex hormone reduction on the response of mouse liver to carcinogens. In a recent experiment, Vesselinovitch (30) showed that gonadectomy in males and females abolished the enhancing effect of partial hepatectomy upon the production of liver tumors by ethylnitrosourea. The effect was more marked in the male. However, the benign tumors, rather than the malignant ones, were affected. This fact should be kept in mind in view of the controversy on the diagnosis of mouse hepatic tumors. In an earlier experiment the same author (31) showed that gonadectomy suppresses the production of hepatic tumors in male mice treated with urethane in the neonatal period and in infancy (6 injections over 18 days), while ovariectomy enhanced the production of this tumor. These two experiments provide results similar to early experiments that demonstrated that removal of sex hormones reduced the spontaneous development of tumors. Note that little attention has been given to the role of other hormones in mouse liver tumor development. Of course, there are data from the rat that indicate that hypophysectomy and thyroidectomy reduce or abolish the hepatocarcinogenic effects of 2-acetylaminofluorene or 2-aminofluorene. However, no comparable mouse experiments have been published (32).

Obviously, systemic studies with a variety of hepatic carcinogens are needed to fully understand the role of hormones on tumor induction by chemicals. In particular, it would be of great assistance if it were possible to determine whether hormones radically affect the metabolism of carcinogens and whether they affect the stage of initiation or promotion.

If there is uncertainty concerning the role that diet and hormones play in the production of hepatic tumors by carcinogens, there is an equal degree of uncertainty concerning the role of genetic factors in altering the response to hepatic carcinogens. Generally speaking, one tends to find a greater response to carcinogens in strains possessing a high natural incidence than in strains with a

# GENETIC, HORMONAL, AND DIETARY FACTORS

low incidence of hepatic tumors. Comparisons are difficult, however, because of differences in dose, duration of treatment, and the numbers of animals on test in the various experiments (11).

The crucial issue is not the absolute incidence of tumors but whether there is a clear difference in the incidence of tumors between test and control. According to my statistician friends, there is no guarantee that this will be the case when the tumor incidence in controls is high. Thus there is no advantage in looking for a strain with a high natural incidence of tumors.

Do these considerations help us in assessing the data from a carcinogenicity test in the mouse? In my view, they show that a mere count of tumors in test and control animals, coupled perhaps with a test of statistical significance, is not adequate. It is imperative to consider the historical data on hepatic tumor incidence and to judge the adequacy of the test data against the background incidence for that particular strain. Further, it is important to look more critically at the biological activity of the material one is asked to test for carcinogenicity in the mouse. It may be a chemical with some hormone-like activity or a food substitute, e.g., single-cell protein. Any increase in tumor incidence in a carcinogenicity test where a protein food is being tested is likely to be a reflection of the high protein intake rather than a direct carcinogenic effect. Much the same attitude could be taken if the compound possessed hormonal activity that promotes growth (e.g., testosterone-like). On the other hand, if the hormone-like activity is one that inhibits tumor production, e.g., estrogen, it could mask a carcinogenic effect.

Above all, the fact that so many factors may operate to confound the result should prevent us from making a precipitate decision about the carcinogenic activity of the substance tested if it induces only hepatic neoplasia in the mouse. Clues that may help us to eliminate any confounding factors and any other data on the biological activity of the substance under test should be sought and carefully considered along with the mouse hepatoma data.

Perhaps the factors mentioned should receive greater consideration before conducting the test. If it seems likely that the test substance in some way resembles proteins, fats, or hormones, then we may well decide not to carry out the carcinogenicity test in the mouse. Alternatively, if, because of regulatory requirements, this sensible course of action cannot be taken, one could build safeguards into the experimental protocol by including appropriate controls. For example, if a protein food is being tested, then a group treated with casein should be included.

Knowledge of the variability of the natural incidence of hepatoma in various strains will also assist in selecting mice from a strain with an appropriately low incidence. It is difficult to define what an appropriately low incidence is, but, in my view, strains with a natural incidence of 10% and over should be avoided.

In conclusion, the natural incidence of mouse hepatoma can be influenced by a number of factors that may interfere with the interpretation of carcinogenicity tests. We have mentioned three but there are others, e.g., tissue injury, germ-free status, and even single caging. Assessment of results from such tests should therefore be made after a careful and prudent search for clues that could lead to an identification of confounding factors and after careful consideration of other relevant biological data.

REFERENCES

1. Tomatis L, Partensky C, Montesano R: The predictive value of mouse liver tumor induction in carcinogenicity testing - A literature survey. Int J Cancer 12:1, 1973.

2. Ward JM, Griesemer RA, Weisburger EK: The mouse liver tumor as an endpoint in carcinogenesis testing. Toxicol Appl Pharmacol 51:389, 1979.

3. Tannenbaum A: The initiation and growth of tumors. Introduction: I. Effects of under-feeding. Am J Cancer 38:335, 1940.

4. Tannenbaum A: Nutrition and cancer. In: The Physiology of Cancer, 2nd ed., edited by H Fishman. New York: Hoeber-Harper, 1959.

5. Tucker MJ: The effect of long-term food restriction on tumors in rodents. Int J Cancer 23:803, 1979.

6. Conybeare G: Effect of quality and quantity of diet on survival and tumor incidence in outbred Swiss mice. Food Cosmet Toxicol 18:65, 1980.

7. Tannenbaum A, Silverstone H: The genesis and growth of tumors IV. Effects of varying the proportion of protein (casein) in the diet. Cancer Res 9:162, 1949.

8. Tannenbaum A: The genesis and growth of tumors III. Effects of a high-fat diet. Cancer Res 2:468, 1942.

9. Gellatly JBM: The natural history of hepatic parenchymal nodule formation in a colony of C57BL mice with reference to the effect of diet. In: Mouse Hepatic Neoplasia, edited by WH Butler, PM Newberne. New York: Elsevier, 1975.

10. Peraino C, Fry RJM, Staffeldt E: Enhancement of spontaneous hepatic tumorigenesis in C3H mice by dietary phenobarbital. J Natl Cancer Inst 51:1349, 1973.

11. Grasso P, Hardy J: Strain difference in natural incidence and response to carcinogenesis. In: Mouse Hepatic Neoplasia, edited by WH Butler, PM Newberne. New York: Elsevier, 1975.

12. Grasso P, Crampton RF, Hooson J: The mouse and carcinogenicity testing. Carshalton, Surrey, England: The British Industrial Biological Research Association, 1977.

13. Shimkin MB, Stoner GD: Lung tumors in mice. Application to carcinogenesis bioassay. Adv Cancer Res 21:1, 1975.

14. Andervont HB: Studies on the occurrence of spontaneous hepatomas in mice of strains C3H and CBA. J Natl Cancer Inst 11:581, 1950.

15. Heston WE: Post-hearing comments. Re: The proposed regulations of toxic substances posing a potential occupational carcinogenic risk to humans. OSHA Docket No. 090, 1978.

16. Heston WE, Vlakakis G: Influence of the Ay-gene on mammary-gland tumors, hepatomas and normal growth in mice. J Natl Cancer Inst 26:969, 1961.

17. Jull JW: Endocrine aspects of carcinogenesis. In: Chemical Carcinogenesis, edited by CE Searle. Washington, DC: American Chemical Society monograph 173, 1976.

18. Ivanova-Chemishanska L, Markov DV, Milanov S, Strashmirov DD, Dashev GI, Chemishanski GA: Effect of subacute oral administration of zinc ethylenebis-(dithiocarbamate) on the thyroid gland and adenohypophysis of the rat. Food Cosmet Toxicol 13:445, 1975.

19. Desser-Wiest L: Stimulation of DNA synthesis in rat liver by adrenalectomy. J Endocrinol 60:315, 1974.

20. Desser-Wiest L: Promotion of liver tumors by steroid hormones. J Toxicol Environ Health 5:203, 1979.

21. Burns EL, Schenken JR: Spontaneous primary hepatomas in mice of strain C3H. A study of incidence, sex distribution and morbid anatomy. Am J Cancer 39:25, 1940.

22. Burns EL, Schenken JR: Spontaneous primary hepatomas in mice of strain C3H. The influence of breeding on their incidence. Cancer Res 3:691, 1943.

23. Murphy ED: Characteristic tumors. In: Biology of the Laboratory Mouse, edited by EL Green, Chap. 27. New York: McGraw-Hill, 1966.

24. Miller EW, Pybus FC: The effect of oestrone in mice of three inbred strains, with special reference to the mammary glands. J Pathol Bacteriol 54:155, 1942.

25. Schenken JR, Burns EL: Spontaneous primary hepatomas in mice of strain C3H. The effect of estrogens and testosterone propionate on their incidence. Cancer Res 3:693, 1943.

26. Agnew LRC, Gardner WV: The incidence of spontaneous hepatomas in C3H, C3H (low milk factor) and CBA mice and the effect of estrogen and androgen on the occurrence of these tumors in C3H mice. Cancer Res 12:757, 1952.

27. Carriere R: The growth of liver parenchymal nuclei and its endocrine regulation. Int Rev Cytol 25:201, 1969.

28. Innes JRM, Ulland BM, Valerio MG, Petrucelli L, Fishbein L, Hart ER, Pallotta AJ, Bates RR, Falk HL, Gart JJ, Klein M, Mitchell J, Peters J: Bioassay of pesticides and industrial chemicals for tumorigenicity in mice: A preliminary note. J Natl Cancer Inst 42:1101-1114, 1969.

29. Kirby AHM: Studies in carcinogenesis with azo compounds. I. The action of four azo dyes in mixed and pure strain mice. Cancer Res 5:673, 1945.

30. Vesselinovitch SD, Itze L, Mihailovich N, Rao KVN: Modifying role of partial hepatectomy and gonadectomy in ethylnitrosourea-induced hepatocarcinogenesis. Cancer Res 40:1538, 1980.

31. Vesselinovitch SD, Mihailovich N: The effect of gonadectomy on the development of hepatomas induced by urethan. Cancer Res 27:1788, 1967.

32. Warwick GP: Metabolism of liver carcinogens and other factors influencing liver cancer induction. In: Liver Cancer, p. 121. IARC Proceedings of a Working Conference held at the Chester Beatty Res. Inst. London, England June 30 to July 3, 1969.

33. Heston WE: Complete inhibition of occurrence of spontaneous hepatomas in highly susceptible (C3H x YBR) $F_1$ male mice by hypophysectomy. J Natl Cancer Inst 31:467, 1963.

Chapter 5

# Kinetics of Induction and Growth of Basophilic Foci and Development of Hepatocellular Carcinoma by Diethylnitrosamine in the Infant Mouse

**Stan D. Vesselinovitch and N. Mihailovich**

INTRODUCTION

The main objective of characterizing the development of chemically induced hepatocellular carcinomas in rodents is to identify the early-appearing hepatocellular population(s) that may represent a progenitor(s) of the hepatocellular carcinomas. In the rat, the emergence of chemically induced hepatocellular carcinomas is preceded by a variety of histochemically characterized hepatocellular foci and hyperplastic nodules (1-5). A relationship between these lesions and carcinomas has been established on the basis of their physical association (6-8), similar histochemical feature(s) (1,2,6,9), and growth kinetics (10). According to Farber and Cameron (11), however, only approximately 5 - 10% of the original hyperplastic nodules persist. In turn, only a very small fraction of those putative precursors may acquire characteristics of hepatocellular carcinomas.

The histogenesis of the hepatocellular carcinoma in the mouse, however, has not been investigated systematically to date. Also, the classification of various nodular liver lesions varies from one laboratory to another. This is mainly due to the lack of studies correlating the structural and morphologic characteristics of well-defined lesions with their biologic behavior both in the primary and secondary hosts.

During the last decade, our laboratory has been engaged in studies with the objective of developing a morphologic classification that would pay due attention to the biology of the focal and nodular liver lesions. Table 1 summarizes the diagnostic

---

This work was supported in part by the National Center for Toxicological Research, Jefferson, Arkansas, under contract 222-76-2004 (C), and by the National Cancer Institute under grants CA-25549 and CA-25522.

TABLE 1. Diagnostic Criteria

HYPERPLASTIC NODULE

Expansive growth
Normal liver histologic structure
Central veins and portal triads present

HEPATOCELLULAR ADENOMA

Expansive growth
One-to-two-cell-thick liver plates showing slight to moderate convolution
Absence of central veins and portal triads
Normal hepatocellular morphology

HEPATOCELLULAR CARCINOMA

Invasive growth
Absence of hepatic landmarks
Multicellular plate or sheet-like configurations
Broad sinusoid-like spaces
Hepatocellular morphology ranging from well differentiated to anaplastic

---

criteria that were developed in the course of these studies. Detailed morphologic documentation can be found elsewhere (12). Table 2 gives the summary data of extensive transplantation studies (13) in which viable cell suspensions from primary nodular lesions, measuring 10-15 mm in diameter, were injected subcutaneously into isogeneic recipients (14). Data show that only neoplastic lesions - hepatocellular adenomas and carcinomas - grew successfully upon transplantation. The rate of transplantability increased and the latent periods decreased as the morphology of transplanted lesions changed from benign to well-differentiated to anaplastic. The hepatocellular adenomas grew upon transplantation in 18% of the cases that became manifest after a long latency (65 weeks). In contrast, moderately differentiated to anaplastic hepatocellular carcinomas grew upon transplantation in 67% of the cases, demonstrating their aggressive growth pattern after an average latent period of 5 weeks. The hyperplastic nodules grew occasionally but only if induced by protracted carcinogenic treatment. Even in these cases, they regressed shortly after transplantation. The biologic behavior of the hepatocellular nodular lesions, as observed in the primary and secondary hosts, is summarized in Table 3. Note that distant metastases in the primary hosts and the invasive and metastatic growths in the secondary hosts were seen exclusively following transplantation of hepatocellular carcinomas. These studies gave credence for the classification presented in Table 1. Thus, it is obvious that any morphologic classification that fails to take into consideration the biologic behavior of the lesions

TABLE 2. Comparative Transplantability of Induced Hepatocellular Nodular Lesions[a]

| Morphology of the primary nodular lesions | Number of transplanted primary liver nodules | Total number of recipients | Transplantability Ratio | Transplantability Percent | Average latent period (weeks)[b] |
|---|---|---|---|---|---|
| Hyperplastic nodule | 30 | 150 | 0/30 | 0 | |
| Adenoma | 60 | 300 | 11/60 | 18 | 65 |
| Well-differentiated carcinoma | 35 | 172 | 18/35 | 51 | 19 |
| Moderately-differentiated to anaplastic carcinoma | 15 | 78 | 10/15 | 67 | 5 |

[a] The following agents were used in the induction of lesions: ethylnitrosourea, diethylnitrosamine, benzo[a]pyrene, benzidine, dieldrin, and DDT.
[b] Time between transplantation and clinical identification of subcutaneous growth.

TABLE 3. Biologic Behavior of Hepatocellular Nodular Lesions

|  | Hyperplasia | Benign neoplasia | Malignant neoplasia |
|---|---|---|---|
|  |  | Primary host |  |
| Cause | Evident and directly related | Not immediately evident | Not immediately evident |
| Type of growth | Expansion | Expansion | Invasion |
| Distant metastases | No | No | Yes |
|  |  | Secondary host |  |
| Transplantation |  |  |  |
| Take | Yes | Yes | Yes |
| Growth | Some | Yes | Yes |
| Expansive | Yes | Yes | No |
| Invasive | No | No | Yes |
| Metastases | None | None | Yes |
| Rate | Limited | Slow | Moderate-fast |

and the kinetics of their development can not be considered as morphologically valid and so becomes misleading.

In order to extend and complement those studies, we began to explore morphologic, tinctorial, and histochemical characteristics of the cell population(s) that would emerge soon after the administration of graded dosages of diethylnitrosamine (DEN) in the infant mouse model, which was developed and defined in our laboratory (15-20). The infant mouse has shown high susceptibility to hepatocarcinogenesis following single administration of DEN due to an optimal deethylating capability of liver (15) and a high, concurrent DNA and cellular replication (17).

MATERIALS AND METHODS

Young C57BL/6J virgin females and C3HeB/FeJ $F_1$ males were purchased from Jackson Laboratories, Bar Harbor, Maine. When 8-10 weeks of age, they were bred, 3 females to 1 male. Pregnant females were housed singly in plastic cages. Delivery dates and litter sizes were recorded twice daily, and the 15-day-old B6C3F1 offspring were injected ip with DEN dissolved in saline. The amounts of administered DEN is specified in the tables. Groups of 8-20 DEN-treated and control mice were killed at 10, 20, 30, 40, 50, 60, 70, 80, and 90 weeks following administration of the carcinogen, unless otherwise specified. The mice were fed Teklad 4% mouse food *ad libitum* (Teklad Standard Diets, Winfield, Iowa). They were kept in a temperature-controlled laboratory ($21^{\circ}C$) with a 12-hour light-dark cycle. Sections of liver, measuring about 1 mm in thickness, were taken through the longest axis of each lobe and were either fixed in buffered formalin for paraffin embedding or frozen for cryostat sectioning. Paraffin sections were cut at 5 μm thickness and were stained with hematoxylin and eosin with and without prior incubation in diastase (21). Cryostat sections were stained for glucose-6-phosphatase (22).

RESULTS

Induction of Basophilic, Glucose-6-Phosphatase-Deficient Foci

The earliest identified focal changes (10 weeks) were characterized by an increase in cytoplasmic basophilia (high RNA content) and by the absence of glucose-6-phosphatase (20). These homogeneous-appearing basophilic foci were replaced after the 20th week by more heterogenous lesions. They were characterized by variability of size (average radial length 126 and 318 μm, respectively), diversification in tinctorial characteristics of cells, and the regional presence of mitotic figures. Thus, in the larger foci, clear, acidophilic, and mixed cell populations became apparent in the central part of the lesion and often appeared hydropic, lipid laden, and even necrotic. The basophilic cells were usually distributed along the outer region of the foci. The mitotic figures, when present, were almost always observed within the basophilic cell population. In time, individual foci became conspicuous by their excessive size, not representing an obvious continuum with more homogeneous small foci. The large size indicated that their growth capabilities differed from the majority of the other foci.

Some of the small basophilic foci, which were adjacent to the central and intercalated veins, showed a tendency to "shed" occasional cells into vascular lumina. In addition, vascular herniation of basophilic hepatocytes was observed. These structures would extend up to 500 μm beyond the point of vascular entry and were usually separated from the intravascular space by the sinusoidal lining. At a farther distance, however, free-floating cells were observed within the vessel lumina. These cells eventually showed a loss of nuclei and cellular lysis. Systematic serial sections of lungs of such animals, however, did not reveal the presence of any hepatic cell emboli, suggesting that these lesions were hyperplastic rather than neoplastic.

Focal lesions were evaluated morphometrically utilizing the Hewlett-Packard digitizer, which was connected with both a Leitz microscope and a computer unit. The average number of focal lesions per cubic centimeter of liver was evaluated using the formula developed by Fullman (23):

$$\underline{N} = \frac{1/\underline{r}_1 + 1/\underline{r}_2 + 1/\underline{r}_3 \ldots 1/\underline{r}_n}{\pi \underline{A}}$$

where $\underline{N}$ is the number of foci/cm$^3$; $\underline{r}_1$, $\underline{r}_2$, $\underline{r}_3$, and $\underline{r}_n$ are the radii (cm) of the observed focal cross sections; and $\underline{A}$ is the area (cm$^2$) of the liver section evaluated. The number of foci per liver was calculated by multiplying foci/cm$^3$ by the weight of the liver in grams. Fullman's method has been used recently by Moore et al, (24) and Campbell et al, (25), although other quantitative methods are available (5,26).

Table 4 gives the experimental protocol and information on the number of lesions per liver and the number of mitotic figures observed per liver cross section at 40 weeks. The data indicate a positive relationship between the DEN dose and the number of focal lesions and mitotic figures. Because the mitotic rate (column 6) stayed constant regardless of DEN dose, it was assumed that the dose-dependent increase in the mitotic frequencies resulted from an increase in the number of mitotically active foci (column 7).

Dose Versus Transformation Probability

Table 5 lists the estimated frequencies for each of four morphologic entities: basophilic foci, hyperplastic nodules, hepatocellular adenomas, and hepatocellular carcinomas for each DEN dose. The basophilic foci occurred most frequently, while the hepatocellular carcinomas rarely developed. The transformation probabilities of any given lesion regardless of DEN dose, are listed in the last row. To calculate these probabilities, it was estimated that at

TABLE 4. Dose-Dependent Variation in Multiplicity of Nodular Liver Lesions and Mitotic Activity at 40 Weeks Following DEN Treatment

| Group | Number Of mice | Treatment[a] (μg/g BW) | Number of lesions/liver Average ± SE | Average number of mitoses/cross section[b,c] | Mitotic rate | Average number of lesions with mitotic figures/cross section[d,e] |
|---|---|---|---|---|---|---|
| 1 | 8 | 0.000 | 0 | 0.0 | 0.0 | 0.00 |
| 2 | 8 | 0.625 | 112 ± 35 | 0.5 | 0.5 | 0.25 |
| 3 | 8 | 1.250 | 262 ± 52 | 12.0 | 3.3 | 3.37 |
| 4 | 8 | 2.500 | 468 ± 99 | 22.4 | 3.7 | 6.87 |
| 5 | 8 | 5.000 | 1207 ± 110 | 42.5 | 3.2 | 11.75 |

[a] Single ip injection of DEN administered to 15-day-old C57BL/6JxC3HeB/FeJ F$_1$ male mice.
[b] Nodular hepatocellular lesions observed in sets of 8 liver cross sections. Frequencies of observed lesions were adjusted to an area of 1.5 cm$^2$/liver, and the numbers/liver were estimated by Fullman's formula.
[c] Mitotic figures identified in sets of 8 liver cross sections, each 1.5 cm$^2$.
[d] Number of mitotic figures identified per 1000 hepatocytes within nodular lesions.
[e] An area of 1.5 cm$^2$.

TABLE 5. Estimated Frequencies of Basophilic Foci, Hyperplastic Nodules, Adenomas, and Carcinomas per Liver, 40 Weeks Following DEN Treatment

| DEN dose ($\mu$g/g) | Basophilic foci A | Basophilic foci B | Hyperplastic nodules | Adenomas | Carcinomas |
|---|---|---|---|---|---|
| 0.625 | 99 | 13 | 0 | 0 | 0 |
| 1.250 | 196 | 13 | 53 | 0 | 0 |
| 2.500 | 378 | 27 | 53 | 10 | 0 |
| 5.000 | 946 | 42 | 184 | 35 | 0.25 |
| Total | 1619 | 95 | 290 | 45 | 0.25 |
| Average | 405 | 24 | 72 | 11 | 0.06 |
| Transformation probability | $0.6 \times 10^{-5}$ | $0.4 \times 10^{-6}$ | $1.2 \times 10^{-6}$ | $1.8 \times 10^{-7}$ | $1.0 \times 10^{-9}$ |

B = foci showing intravascular protrusion.

the time of carcinogenic treatment (15 days of age), approximately $6 \times 10^7$ liver cells were at "transformation" risk. Assuming that each of the above lesions originated from a single liver cell, the transformation probability (TP) for each morphologic entity was calculated by dividing the number of estimated lesions per liver ($N_L$) by the number of liver cells at transformation risk ($N_{TR}$):

$$TP = \frac{N_L}{N_{TR}}$$

The transformation probability ranged from $0.7 \times 10^{-5}$, for basophilic foci (A and B combined), to $1 \times 10^{-9}$ for hepatocellular carcinoma, a ratio of 4 orders of magnitude. The transformation probability of hepatocellular adenomas was 2 orders of magnitude higher than that of hepatocellular carcinoma ($1.8 \times 10^{-7}$). It is thus obvious that the transformation probability of the spectrum of liver lesions decreased as the degree of morphologic and biologic deviation from normalcy increased.

The overall shape of the curves depicting numbers of basophilic foci against time was sigmoid, so that, in general, the number of foci increased exponentially. The increase in carcinogenic dose, however, resulted in a shorter latent period. The exponential shape of those curves suggests that cellular growth was the underlying mechanism. An increase in labeling indices and the mitotic activities further substantiated this supposition.

Table 6 presents the effect of dose on the number of basophilic foci per liver at 30 weeks. Note the positive regression of the basophilic foci on DEN dose. Slope (footnote a: $b$ = 54.63) shows that a dose of 1 μg DEN/gm body weight induced 54.6 basophilic foci. The transformation probabilities of basophilic foci are tabulated in the last column. Data show that this probability is directly related to carcinogenic dose, showing a high correlation value ($r$ = 0.997). The slope of this correlation (footnote b) shows that the number of focus-forming cells induced was $0.91/10^6$ liver cells for each μg DEN/gm body weight. At 40 weeks, the dose-response slope indicated that the number of focus-forming cells induced was $3.2/10^6$ liver cells for each μg DEN/gm body weight. This shows a constant dose effect upon a relatively high transformation probability.

Dose Versus Time to 50% Incidences
of Various Morphologic Entities

Additional information on kinetics of hepatocarcinogenesis can be obtained by correlating the time (weeks) in which 50% of the animals would bear basophilic foci with the dose of DEN. This is referred to as $t_{50\%}$ incidence. Such a correlation may offer information regarding the pattern of the accelerating effect of DEN dose upon the induction of a specific morphologic entity. Table 7 illustrates regression of $t_{50\%}$ incidence of basophilic foci on DEN dose. One may observe a high correlation coefficient ($r$ = -0.9950; double

TABLE 6. Effect of DEN Dose on The Frequency of Basophilic Foci 30 Weeks Following Treatment

| Group | DEN dose ($\mu$g/g) | Number of foci[a] per liver (X $\pm$ SE) | Transformation probabilities of basophilic foci[b] |
|---|---|---|---|
| 1 | 0.000 | 0.0 $\pm$ 0.00 | 0.000 x $10^{-6}$ |
| 2 | 0.625 | 24.4 $\pm$ 3.42 | 0.407 x $10^{-6}$ |
| 3 | 1.250 | 63.0 $\pm$ 5.23 | 1.050 x $10^{-6}$ |
| 4 | 2.500 | 114.0 $\pm$ 9.71 | 1.900 x $10^{-6}$ |
| 5 | 5.000 | 266.0 $\pm$ 16.76 | 4.433 x $10^{-6}$ |

[a] $r = 0.977$; $a = -11.919$; $b = 54.63$.
[b] $r = 0.997$; $a = -0.186$; $b = 0.91 \times 10^{-6}$.

logarithmic scale) and an excellent agreement between the observed (column 2) and expected (column 5) $t_{50\%}$ values. Data show a constant rate of acceleration of development of basophilic foci corresponding to the dose of DEN.

The animals killed later in life developed hepatocellular carcinomas in addition to the other focal lesions. Their cumulative incidences are presented in Fig. 1. This family of dose-response curves illustrates two points: first, each dose of DEN-induced

TABLE 7. Regression of Time to 50% Incidence of Basophilic Foci on DEN Dose

| Dose ($\mu$g/g) | Observed time (weeks) to 50% | log (dose) | log (time) | Expected time (weeks) to 50% |
|---|---|---|---|---|
| 0.625 | 26.5 | -.204 | 1.42 | 26.46 |
| 1.250 | 20.0 | .097 | 1.30 | 20.46 |
| 2.500 | 16.5 | .398 | 1.22 | 15.83 |
| 5.000 | 12.0 | .699 | 1.08 | 12.24 |

$r = -0.9950$; $a = 1.3454$; $b = -0.3654$; $n = 2.73$.

%
100 —

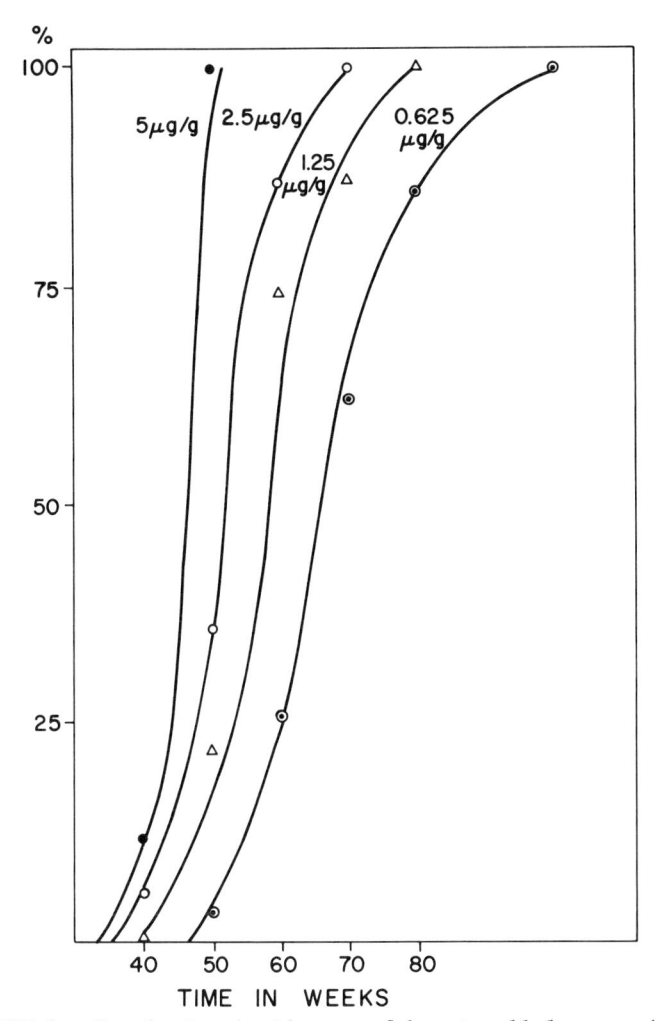

5μg/g  2.5μg/g   0.625 μg/g
         1.25 μg/g

FIGURE 1. Cumulative incidences of hepatocellular carcinomas following single treatment. Specified dosages of diethylnitrosamine (DEN) were injected intraperitoneally into 15-day-old B6C3F$_1$ male mice. Time was measured after carcinogenic treatment.

hepatocellular carcinomas in all animals, and second, the increase in carcinogenic dose accelerated the development of carcinomas.

Table 8 lists regression of time to 50% incidence of hepatocellular carcinomas on DEN dose. Note again the high correlation coefficient ($r$ = -0.999) and an excellent agreement between the observed (column 2) and the expected (column 5) time to 50% incidence of hepatocellular carcinoma. These data illustrate a constant rate of acceleration for the induction of hepatocellular carcinoma relating to the DEN dose. Since at $t_{50\%}$ incidence of hepatocellular carcinoma there

TABLE 8. Regression of Time to 50% Incidence of Hepatocellular Carcinoma on DEN Dose

| Dose (µg/g) | Observed time (week) to 50% | log (dose) | log (time) | Expected time (weeks) to 50% |
|---|---|---|---|---|
| 0.625 | 65.9 | -.204 | 1.819 | 65.82 |
| 1.250 | 58.9 | .097 | 1.770 | 58.74 |
| 2.500 | 52.0 | .399 | 1.716 | 52.42 |
| 5.000 | 46.9 | .699 | 1.672 | 46.08 |

$\underline{r}$ = -0.999; $\underline{a}$ = 1.785; $\underline{b}$ = -0.164.

is on an average 0.5 carcinoma per carcinoma-bearing animal, the estimated transformation probability is $0.8 \times 10^{-8}$ (0.5 carcinoma-forming cell/$6 \times 10^7$ liver cells at risk). Accordingly, time to 50% incidence of hepatocellular carcinoma ($\underline{t}_{50\%Ca}$) could be designated also as time to $10^{-8}$ transformation probability of carcinoma.

The comparative relationship between the DEN dose and the time needed for 50% incidence of basophilic foci and hepatocellular carcinoma is illustrated in Fig. 2. The respective $\underline{t}_{50\%}$ values were plotted against the carcinogenic dose on a double logarithmic scale. Such a presentation of data is based on Druckrey's observation that, within a certain dose range, there is a constant relationship between the daily dose of carcinogen ($\underline{d}$) and the time of administration ($\underline{t}$) required for carcinoma development ($d\underline{t}^n = \underline{k}$) (4). This figure illustrates the constancy of the dose-time relationship for both basophilic foci and hepatocellular carcinomas. The exponents on time ($\underline{n}$), however, differed, being 2.6 and 5.7 for these two morphologic endpoints, respectively. Table 9 tabulates dose-dependent time to 50% incidence of all four morphologic entities: basophilic foci, hyperplastic nodules, hepatocellular adenomas, and hepatocellular carcinomas. The first two endpoints have a similarly low value of $\underline{n}$ (2.6 and 2.7 respectively). The $\underline{n}$ value for hepatocellular adenoma was of an intermediate magnitude ($\underline{n}$ = 3.4).

The crucial distinction has to be made between the designated meanings of dose ($\underline{d}$) and time ($\underline{t}$) in studies carried out originally by Druckrey (27) and those presented here. In the present case, the formula $d\underline{t}^n = \underline{k}$ applies to the single-dose carcinogenesis, so that the time factor ($\underline{t}$) is free of any carcinogenic treatment. According to Druckrey, the numeric value of $\underline{n}$ is an indicator of the carcinogenic potency of the agent. Since we compared four distinct morphologic entities in this study, the numeric value of $\underline{n}$ may indicate the degree by which they deviated from the normal cells population. The interpretation by Emmelot and Scherer (28)

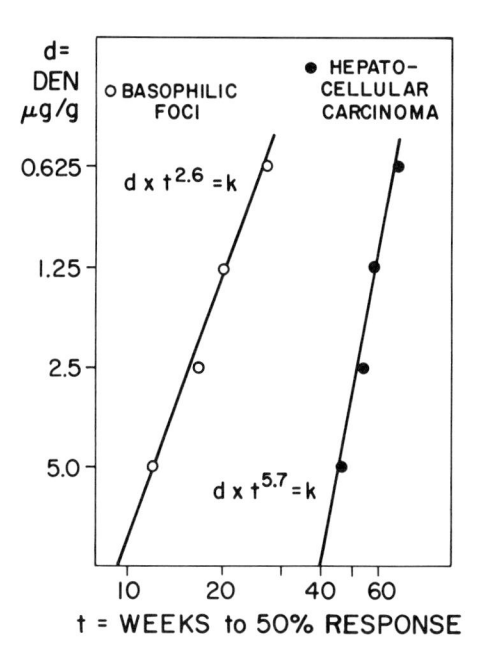

FIGURE 2. Regression of time to 50% incidence of mice bearing basophilic foci (O) or hepatocellular carcinomas (●) on the dose of diethylnitrosamine (DEN) (double logarithmic plot). Mice were B6C3F$_1$ injected once ip at 15 days of age.

of the meaning of $\underline{n}$ value may be more appropriate here. These authors suggested that the value of $\underline{n}$ may denote the rate of proliferation of the intermediate cell populations in the sequence from normal to tumor cells. Their position assumes that once the neoplastic process has been set in motion, the time required for neoplastic expression will depend upon the number of cycles (and their duration) that is necessary for the expression of commitment of each initiated-cell subpopulation. If so, it stands to reason that the subpopulation of the cells that were committed to grow out into the basophilic foci will require fewer cell cycles for their expression ($\underline{t}^{2.6}$) than the subpopulation that has been committed to the hepatocellular carcinomas ($\underline{t}^{5.7}$).

Dose Versus Growth, Mitotic Activity and
Development of Hepatocellular Carcinoma

The weight of the livers was one of the measurable endpoints taken in the course of the hepatocarcinogenesis studies. As the nodular lesions emerged, the weight of the liver increased in relation to the number and size of nodular lesions. A plot of liver weights against time for each DEN dose gave a family of curves similar in

TABLE 9. Dose-Dependent Time to 50% Incidence for Morphologic Endpoints Following Single Administration of Diethylnitrosamine

| Morphologic endpoints | Dose of DEN (μg/gm body weight)[a] | | | | n Values (tangens) |
|---|---|---|---|---|---|
| | 5.0 | 2.5 | 1.25 | 0.625 | |
| Basophilic foci | 12.0[b] | 16.5 | 20.0 | 26.5 | 2.6 |
| Hyperplastic nodules | 21.5 | 28.5 | 38.0 | 46.5 | 2.7 |
| Hepatocellular adenomas | 30.0 | 35.0 | 43.0 | 55.0 | 3.4 |
| Hepatocellular carcinomas | 46.9 | 52.0 | 58.9 | 65.9 | 5.7 |

[a] Specified dosages were administered intraperitoneally to 15-day-old B6C3F$_1$ male mice.
[b] Time to 50% incidence (weeks).

shape and relationship to curves obtained by plotting cumulative incidences of hepatocellular carcinomas on DEN dose (Fig. 1). It was found that the time at which the average liver weights were 2.8 gm ($t_{2.8gm}$) was similar to the time at which 50% of the animals developed hepatocellular carcinomas ($t_{50\%}$). Table 10 shows this parallelism, indicating a casual relationship between the carcinogenic dose, the nodular growth, and the carcinoma development. This relationship is further substantiated by the high, positive correlation between the dose of DEN and the average number of mitotic figures observed at 40 weeks (Fig. 3). It is apparent that single DEN treatments set in motion processes resulting in a dose-dependent cell replication. Table 11 presents the regression of time to 50% incidence of hepatocellular carcinoma (dependent variable) on the mitotic activity (independent variable) on the logarithmic scale. Since the mitotic activity was estimated at 40 weeks following DEN treatment, the time to 50% incidence of carcinoma was measured from that point in time. From the slope (b in Table 11), we calculated the n value and found the constancy of product of mitotic activity (m) and time to 50% incidence of carcinoma ($mt^{1.01}$ = k). Since the n value did not differ from unity, it may be concluded that at 40 weeks the speed of development of carcinoma was dependent on the proliferating rate of cancer-committed clones. This gives further credence to the suggestions made by others (5,29) that proliferation of precancerous clones represents the main factor in determining the time effect (n value).

The differential in growth potential of various morphologic entities is also illustrated in Table 12. It lists, for each morphologic entity, the radial size, the number of cells, and the required

TABLE 10. Parallelism Between Liver Growth and Carcinoma Development and Their Dependence upon Carcinogenic Dose

| DEN dose | Time to 2.8 gm liver weight | Time to 50% hepatocellular carcinoma |
|---|---|---|
| 0.625 | 63.4 | 65.9 |
| 1.250 | 59.2 | 58.9 |
| 2.500 | 53.6 | 52.0 |
| 5.000 | 46.8 | 46.9 |

number of cell cycles. The numbers were estimated by assuming that (1) each lesion originated from a single cell, (2) all cells underwent division, and (3) no cells were lost. Data presented in this table were obtained from mice that were killed 36 weeks following administration of 10 µg DEN/gm body weight at 15 days of age. There were five distinct radial size groups: two for basophilic foci (focus$_1$ and focus$_2$) and one each for hyperplastic nodules, hepatocellular adenomas, and hepatocellular carcinomas. The differential growth of these morphologic entities is also illustrated by their estimated number of cells and the required number of cell cycles. The highest growth (size) was seen in the case of hepatocellular carcinomas.

It is essential to inspect the last column in Table 12, which gives percentages for each morphologic entity. Although times to 50% incidence of basophilic foci, hyperplastic nodules, and hepatocellular adenomas were 10, 20, and 28 weeks respectively, the majority of the lesions (69.6%) were still basophilic foci. Thus, sequential emergence of various lesions does not imply their total sequential replacement. It is for this reason that the other focal lesions were seen concurrently with carcinomas.

DISCUSSION

Presented data showed that single low, nontoxic dose levels of DEN induced, in infant B6C3F$_1$ male mice, a spectrum of well-defined morphologic entities including hepatocellular carcinomas. Treatment of young adults (42 days of age) with up to 50 µg DEN/gm body weight failed to induce any nodular lesions within the same observational period (personal observation). This points out the significance of cell replication at the time of carcinogenic treatment as the crucial factor in hepatocarcinogenesis. This observation is in agreement with studies carried out by Pitot and Sirica (30), Cayama et al. (31), Emmelot and Scherer (28), and Rabes et al.

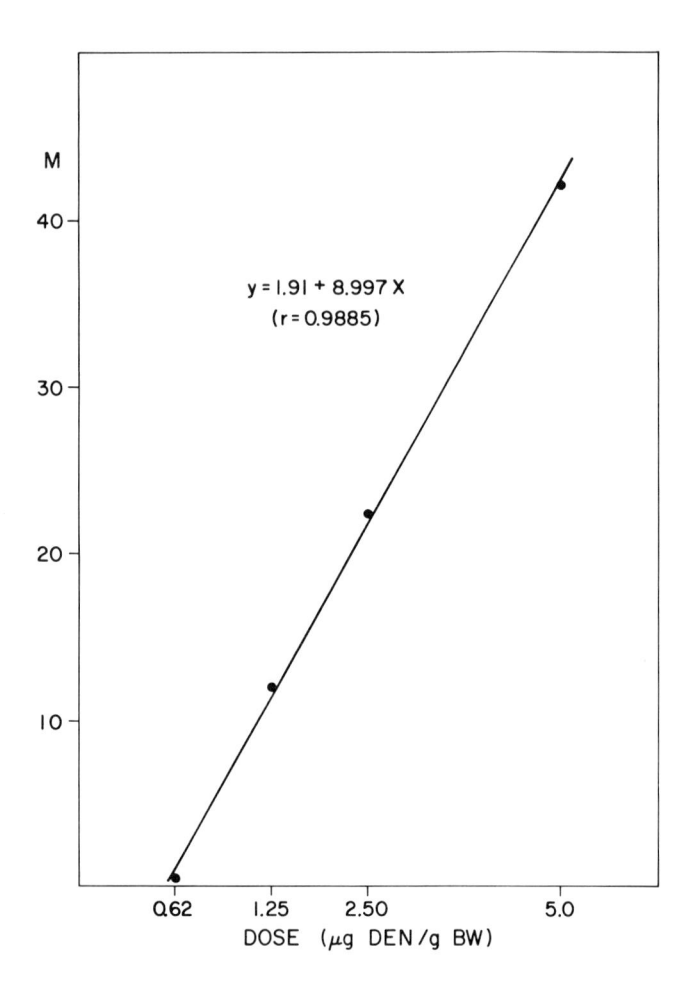

FIGURE 3. Regression of mitotic activity on the dose of diethylnitrosamine (DEN). DEN was injected intraperitoneally into 15-day-old B6C3F$_1$ male mice. The average numbers of mitotic figures (M) sections were evaluated at 40 weeks.

(32,33) in rats. These authors observed an enhancement of the emergence of enzyme deficient islands and acceleration of development of hepatocellular carcinoma by partial hepatectomy. It is likely that the action of carcinogen during DNA synthesis (34) more effectively causes base mispairing and its error-prone repair, leading to more efficient induction of the primary biochemical lesions. The cell replication, following carcinogenic exposure, would lead to "fixation" of the primary lesions, amplification of

TABLE 11. Regression of Time to 50% Hepatocellular Carcinoma on Mitotic Activity

| Average number of mitotic figures[a] | Observed time to 50% hepatocellular carcinoma (weeks)[b] | log (Mitoses) | log ($t_{50\%}$ Hepatocellular carcinoma) | Expected time to 50% hepatocellular carcinoma (weeks) | Statistics |
|---|---|---|---|---|---|
| 12.0 | 18.9 | 1.079 | 1.2764 | 20.9 | $r$ = -0.9870 |
| 22.4 | 12.5 | 1.350 | 1.0969 | 11.3 | $a$ = 2.3853 |
| 42.5 | 6.7 | 1.628 | 0.8260 | 6.0 | $b$ = -0.9876 |
| 71.3 | 3.2 | 1.853 | 0.5051 | 3.6 | $n$ = 1.0125 |

[a]Observed in 1.5-cm$^2$ cross sections of liver 40 weeks after DEN treatment.
[b]After 40th week.

TABLE 12. Size, Number of Cells, and Percent of Lesions Induced per Liver[a] by Diethylnitrosamine[b]

| Morphologic endpoints | Radius ± SD (μm) | Estimated number Cells/lesion | Estimated number Cell cycles/lesion | Percent induced lesions |
|---|---|---|---|---|
| Basophilic focus$_1$ | 126 ± 30 | 1,166 | 10 | 9.6 |
| Basophilic focus$_2$ | 318 ± 76 | 18,708 | 14 | 60.0 |
| Hyperplastic nodule | 504 ± 82 | 74,472 | 16 | 22.4 |
| Hepatocellular adenomas | 1,009 ± 398 | 597,625 | 19 | 7.9 |
| Hepatocellular carcinomas | 2,544 ± 284 | 9,578,708 | 23 | 0.1 |

[a] Animals were killed at 36 weeks.
[b] At 10 μg/gm body weight injected intraperitoneally into 15-day-old B6C3F$_1$ male mice.

the initiated cells (35,36), and an accelerated emergence of the clones of the initiated cells.

It is tempting to consider the basophilic, glucose-6-phosphatase-deficient foci as progenitors of carcinomas. The presented data, however, argue against such a generalization: (1) the ratio between the number of basophilic foci and the number of hepatocellular carcinomas is high (1000 to 1), and (2) the difference in proliferative activity suggests growth potential heterogeneity and emergence of specific cell subpopulations, only a few of which achieve the malignant state while the majority stay as basophilic foci and hyperplastic nodules.

Thus it is possible that the majority of the various nodular lesions may represent fully expressed morphologic entities. This interpretation implies differential commitment between the initiated cells so that they reach, due to limited replicating potential, the state of basophilic foci, hyperplastic nodules, and even benign neoplasia. Only cells possessing unlimited replicating potential are capable of arriving at the malignant state (see Table 12 and related text).

The advantages of the infant mouse model for studying hepatocarcinogenesis are that: (1) a single low-dose carcinogenic treatment can initiate the carcinogenic process; (2) high cellular replication, a factor essential for hepatocarcinogenesis, occurs spontaneously; and (3) various lesions, including hepatocellular carcinomas, develop sequentially and mainly exist simultaneously. The high cellular replication contributes differentially to carcinogenic processes. It apparently enhances the base mispairing and error-prone repair when present at the time of carcinogenic action, and leads to "fixation" of the primary biochemical lesion(s) when it follows carcinogenic treatment. The "fixation" results in cellular initiation, giving origin to specifically committed clones. It is likely that more cell divisions are required for the "fixation" of the biochemical lesion(s) affected by the administration of the lower carcinogenic dose, rather than following the administration of the higher carcinogenic dose. The enhanced replication of the initiated cells shortens the time necessary for the phenotypic expression.

Certain foci that had reached a critical size (approximately 3 mm in diameter) began to manifest increased cellular replication usually restricted to a delineated area. This was followed by the emergence of cell subpopulations that possess replicating autonomy of various degrees. Only those that achieve complete autonomy become neoplastic in character.

SUMMARY AND CONCLUSIONS

The most significant observations in this study are:

1. Induction of four biologically and morphologically distinct entities by a single carcinogenic dose of DEN.

2. Sequential emergence of basophilic foci (18.7 ± 3.0 weeks), hyperplastic nodules (33.6 ± 5.4 weeks), hepatocellular adenomas (40.7 ± 5.4 weeks), and hepatocellular carcinomas (55.9 ± 4.1 weeks).
3. A highly significant difference in the proportion of basophilic foci (82%), hyperplastic nodules (15%), hepatocellular adenomas (2.9%), and hepatocellular carcinomas (0.1%) by $t_{50\%Ca}$ (55.9 weeks).
4. The product of a single dose ($\underline{d}$) of DEN and the time to 50% incidence ($\underline{t}_{50\%}$) is constant ($\underline{k}$) for each morphologic entity: $\underline{dt}^{\underline{n}} = \underline{k}$.
5. The numeric value of exponent $\underline{n}$ increases from 2.6 to 2.7, 3.4, and 5.7 for basophilic foci, hyperplastic nodules, hepatocellular adenomas, or hepatocellular carcinomas, respectively.
6. The product of the average mitotic activity ($\underline{m}$) and the time to 50% incidence of carcinoma ($\underline{t}_{50\% Ca}$) is constant ($\underline{k}$): $\underline{mt}^{1.01} = \underline{k}$.

These observations indicate that: (1) there is a significant difference in the biological commitment between the "initiated" cells, (2) that the majority of the morphologic lesions, observed late in carcinogenesis, represent the fully expressed biologic entities, and (3) that the probability that the early lesions may give rise to hepatocellular carcinoma is low ($\underline{p} < 0.001$). Therefore, one may conclude that the sequential appearance of the above lesions does not imply necessarily morphologic progression, since it could be viewed as representing a series of independent morphologic expressions, each requiring progressively longer latent periods directly related to their biologic complexity. Extended persistence of the early-emerging lesions gives more credence to the latter interpretation.

REFERENCES

1. Gossner W, Friedrich-Freksa H: Histochemische untersuchungen uber die glucose-6-phosphatase in der rattenleber wahrend der kanzerisierung durch nitrosamine. Z Naturforsch [B] 19:862-863, 1964.

2. Kelengayi MMR, Ronchi G, Desmet VJ: Histochemistry of gamma-glutamyl transpeptidase in rat liver during aflatoxin $B_1$-induced carcinogenesis. J Natl Cancer Inst 55:579-588, 1975.

3. Sasaki T, Yoshida T: Experimentelle erzeugung des lebercarcinomas durch futterung mit $\underline{O}$-aminoazotoluol. Virchows Arch [Pathol Anat] 295:175-200, 1935.

4. Schauer A, Kunze E: Enzyme histochemische und autoradiographische untersuchungen wahrend der cancerisierung der rattenleber mit diethylnitrosamin. Z Krebsforsch 70:252-266, 1968.

5. Scherer E, Emmelot P: Kinetics of induction and growth of precancerous liver-cell foci, and liver tumor formation by diethylnitrosamine in the rat. Eur J Cancer 11:689-696, 1975.

6. Farber E: The pathology of experimental liver cell cancer. In: Liver Cell Cancer, edited by HM Cameron, DA Linsell, GP Warwick, pp. 243-257. Amsterdam: Elsevier, 1976.

7. Goldfarb S: A morphological and histochemical study of carcinogenesis of the liver in rats fed 3-methyl-4-dimethylaminoazobenzene. Cancer Res 33:1119-1128, 1973.

8. Pugh TD, Goldfarb S: Quantitative histochemical and autoradiographic studies of hepatocarcinogenesis in rats fed 2-acetylaminofluorene followed by phenobarbital. Cancer Res 38:4450-4457, 1978.

9. Goldfarb S, Zak FG: Role of injury and hyperplasia in the induction of hepatocellular carcinoma. J Am Med Assoc 178:729-731, 1961.

10. Kunz W, Appel KE, Rickart R, Schwarz M, Stockle G: Enhancement and inhibition of carcinogenic effectiveness of nitrosamines. In: Primary Liver Tumors, edited by H Remmer, HM Bolt, T Bannash, H Popper, pp. 261-283. Baltimore: University Park Press, 1978.

11. Farber E, Cameron R: The sequential analysis of cancer development. Adv Cancer Res 31:125-226, 1980.

12. Vesselinovitch SD, Mihailovich N, Rao KVN: Morphology and metastatic nature of induced hepatic nodular lesions in C57BLxC3H $F_1$ mice. Cancer Res 38:2003-2010, 1978.

13. Vesselinovitch SD, Mihailovich N: Characterization of induced mouse nodular liver lesions by isogeneic transplantation. Reg Toxicol Pharmacol, 1982.

14. Kyriazis A, Koka M, Vesselinovitch SD: Metastatic rate of liver tumors induced by diethylnitrosamine in mice. Cancer Res 34:2881-1886, 1974.

15. Rao KVN, Vesselinovitch SD: Age- and sex-associated diethylnitrosamine dealkylation activity of the mouse liver and hepatocarcinogenesis. Cancer Res 33:1625-1627, 1973.

16. Itze L, Vesselinovitch SD, Rao KVN: Estimation of the rate of DNA synthesis in newborn, regenerating, and intact mouse livers. Physiol Bohemoslov 22:457-460, 1973.

17. Itze L, Vesselinovitch SD, Rao KVN: Inter and intra diurnal variations of DNA, RNA and protein synthetic activity in

newborn, infant and young adult mouse livers. Physiol Bohemoslov 25:289-293, 1976.

18. Vesselinovitch SD, Rao KVN, Mihailovich N: Neoplastic response of mouse tissues during perinatal age periods and its significance in chemical carcinogenesis. J Natl Cancer Inst 51:239-250, 1979.

19. Vesselinovitch SD: Infant mouse as a sensitive bioassay system for carcinogenicity of N-nitroso compounds. In: N-Nitroso Compounds: Analysis, Formation and Occurrence, edited by EA Walker, M Castegnaro, L Griciute, M Börzsönyi, pp. 645-655. IARC Scientific Publication no. 31, Lyon, France, 1980.

20. Goldfarb S, Vesselinovitch SD, Pugh TD, Mihailovich N, Koen H, He Y: Tumor progression during diethylnitrosamine (DEN) induced mouse hepatocarcinogenesis. Proc Am Assoc Cancer Res 22:491, 1981.

21. Barka T, Anderson PJ: Histochemistry-Theory, Practice and Bibliography. New York: Harper and Row, 1963.

22. Wachstein M, Meisel E: On the histochemical demonstration of glucose-6-phosphatase. J Histochem Cytochem 4:592, 1956.

23. Fullman RL: Measurement of particle sizes on opaque bodies. Trans AIME 197:447-452, 1953.

24. Moore MR, Drinkwater NR, Miller EC, Miller JA, Pitot HC: Quantitative analysis of the time-dependent development of glucose-6-phosphatase-deficient foci in the livers of mice treated neonatally with diethylnitrosamine. Cancer Res 41:1585-1593, 1981.

25. Campbell HA, Pitot HC, Potter VR, Laishes BA: Application of quantitative stereology to the evaluation of enzyme-altered foci in rat liver. Cancer Res 42:465-472, 1982.

26. Scherer E, Hoffmann M, Emmelot P, Friedrich-Freksa H: Quantitative study on foci of altered liver cells induced in the rat by a single dose of diethylnitrosamine and partial hepatectomy. J Natl Cancer Inst 49:93-106, 1972.

27. Druckrey H: Quantitative Aspects in Chemical Carcinogenesis. From: UICC Monograph Series 7, edited by R Truhaut, pp. 60-78. New York: Springer-Verlag, 1967.

28. Emmelot P, Scherer E: Multi-hit kinetics of tumor formation with special reference to experimental liver and human lung carcinogenesis and some general conclusions. Cancer Res 37:1702-1708, 1977.

29. Scherer E, Emmelot P: Foci of altered liver cells induced by a single dose of diethylnitrosamine and partial hepatectomy:

Their contribution to hepatocarcinogenesis in the rat. Eur J Cancer 11:145-154, 1975.

30. Pitot HC, Sirica AE: The stages of initiation and promotion in hepatocarcinogenesis. Biochim Biophys Acta 605:191-215, 1980.

31. Cayama E, Tsuda H, Sarma DSR, Farber E: Initiation of chemical carcinogenesis requires cell proliferation. Nature (Lond) 275:60-62, 1978.

32. Rabes HM, Scholze P, Jantsch B: Growth kinetics of diethylnitrosamine-induced, enzyme-deficient "preneoplastic" liver cell populations in vivo and in vitro. Cancer Res. 32:2577-2586, 1972.

33. Rabes HM, Szymokowiak R: Cell kinetics of hepatocytes during the preneoplastic period of diethylnitrosamine-induced liver carcinogenesis. Cancer Res 39:1298-1304, 1979.

34. Rabes HM, Kerler R, Wilhelm R, Rode G, Riess H: Alkylation of DNA and RNA by [$^{14}$C]dimethylnitrosamine in hydroxyurea-synchronized regenerating rat liver. Cancer Res 39:4228-4236, 1979.

35. Vesselinovitch SD: Factors modulating response to carcinogenic mutagens. In: Progress in Environmental Mutagenesis, edited by M Alacevic, pp. 281-296. Amsterdam: Elsevier/North Holland Biomedical Press, 1980.

36. Vesselinovitch SD, Itze L, Mihailovich N, Rao KVN: Modifying role of partial hepatectomy and gonadectomy in ethylnitrosourea-induced hepatocarcinogenesis. Cancer Res 40:1538-1542, 1980.

Chapter 6

# Comparison of Morphologic and Biologic Characteristics of Liver Tumors in Control Mice and Mice Treated with 2-Acetylaminofluorene or Benzidine Dihydrochloride

**Charles H. Frith**

INTRODUCTION

A number of investigators have suggested that liver tumors induced with hepatocarcinogens may be morphologically different from those in control mice (1-4). If this were true, it would suggest that the chemical carcinogens are inducing rather than promoting liver tumors. This study attempted to address this question by examining and comparing some of the biological and morphological characteristics of liver tumors in control mice and in mice treated with 2-acetylaminofluorene (2-AAF) or benzidine dihydrochloride.

MATERIALS AND METHODS

This report includes subsets of mice from two larger studies, one of which has been previously reported (5). Livers from 287 control and 144 benzidine-treated $F_1$ hydrid female mice (C57BL/6JfC3Hf/Nctr females x BALB/cStCrlfC3H/Nctr males) and 416 control and 141 2-AAF-treated BALB/c StCrlfC3H/Nctr female mice were evaluated without knowledge of treatment. The BALB/c mice were fed a meal diet (Purina 5010C) containing either 0 or 150 ppm of 2-AAF. The $F_1$ hybrid mice were fed Purina 5010C meal and water containing either 0 or 120 ppm benzidine dihydrochloride <u>ad libitum</u>.

Animals were housed, four per cage, under specific pathogen-free defined flora (SPF-DF) barrier conditions in rooms maintained at 22-24°C and 50 ± 10% relative humidity. At death or moribund sacrifice, each animal was assigned an identification number, and a necropsy was performed. Gross and microscopic findings were collected, as previously described (6,7), on approximately 45 tissues or organs and on grossly visible lesions not included with these organs. Sections of the median lobe of the liver were collected routinely, and gross lesions involving other lobes were also collected for microscopic evaluation. The tissues were fixed in Bouin's solution for 18-24 hours. After fixation, the tissues were trimmed, placed in plastic cassettes for processing on an Autotechnicon on a 4-hour cycle, and embedded in paraffin blocks.

Paraffin sections were cut at 5 µm and stained on an automatic stainer with hematoxylin and eosin (H&E) as described by Frith and Konvicka (6). Liver tumors were classified morphologically according to the classification of Frith and Ward (8).

RESULTS

Biologic and morphologic characteristics between the liver tumors in the control and treated mice included the ratio of single to multiple tumors, mean age of detection, average size, ratio of adenomas to carcinomas, ratio of well to moderately well to poorly differentiated hepatocellular carcinomas, incidence of a prominent trabecular pattern, incidence of metastases, cause of death, tinctorial properties of neoplastic hepatocytes, and size of neoplastic hepatocytes.

Biologic characteristics of both the treated and control groups are presented in Table 1. The incidence of hepatocellular neoplasms was higher in the 2-AAF- and benzidine-treated groups than in the respective controls. The mean age of detection ($\pm$ SD) was also less in the treated groups than in the controls.

Liver tumors were much more likely to be single than multiple in the controls than in the treated mice of both groups. The average largest dimension ($\pm$ SD) of the hepatocellular carcinomas was also greater for the treated than the control mice.

The ratio of hepatocellular carcinomas was slightly increased in the treated mice compared to the controls. The degree of differentiation was also slightly less in the treated than control mice. No metastases occurred in either the control or treated group of the BALB/c mice, but the incidence was 0 and 7.8% in the control and treated $F_1$ hybrids, respectively. Hepatocellular carcinomas resulted in death of the animal; 14.3% in the control and 37.8% in the 2-AAF BALB/c mice and 50 and 88% for the control and benzidine-treated $F_1$ mice, respectively.

Hepatocellular neoplasms were classified as either acidophilic or basophilic, depending upon the predominant tinctorial property of the neoplastic cells within a given hepatocellular tumor (Figs. 1-4). The hepatocellular neoplasms were also classified as to the predominant size of the cells composing a given tumor (Figs. 1-4). Hepatocellular tumors composed of cells similar to or smaller than adjacent normal hepatocytes were classified as being composed of small cells. Hepatocellular tumors composed of cells larger than adjacent normal hepatocytes were classified as being composed of large cells.

Morphologic characteristics are presented in Table 2. The ratio of hepatocellular neoplasms composed of basophilic to acidophilic cells was 56:44 in the control BALB/c mice and 16:84 in the BALB/c mice receiving the 2-AAF. In the $F_1$ hybrid mice, the ratio of basophilic cells to acidophilic cells was 55:45 in the control mice and 18:82 in the benzidine-treated mice.

TABLE 1. Comparison of Biologic Characteristics of Spontaneous and Induced Hepatocellular Neoplasms in Mice

| Treatment and strain | Incidence | Single:multiple (%) | Mean age (d) of detection ± SD | Average size (mm) of greatest dimension ± SD | Ratio of adenomas: carcinomas | Ratio of well: moderately well: poorly differentiated hepatocellular carcinomas | Incidence of metastases (%) | Liver tumor cause of death (%) |
|---|---|---|---|---|---|---|---|---|
| BALB/c controls | 27/416 (6.5)[a] | 63:37 | 815±163 | 10.7±5.7 | 6:21 (22:78) | 16:5:0 (76:24:0) | 0/21 (0) | 3/21 (14.3) |
| BALB/C treated | 45/141 (31.9)[b] | 24:76[b] | 710±103 | 13.2±8.7 | 8:37 (18:82) | 24:13:0 (65:35:0) | 0/37 (0) | 14/37 (37.8)[b] |
| F₁ hybrid controls | 11/287 (3.8) | 56:44 | 961±79 | 11.4±5.4 | 3:8 (27:73) | 5:2:1 (63:25:12) | 0/8 (0) | 4/8 (50) |
| F₁ hybrid treated | 124/144 (86.1)[b] | 3:97[b] | 546±80 | 16.3±8.0 | 9:115 (7:93)[c] | 51:61:3 (44:53:3) | 9/115 (7.8) | 101/115 (88) |

[a] Parenthesis indicate incidence, expressed in percent.
[b] Significantly different from the corresponding control ($p < .05$) utilizing a one-sided test.

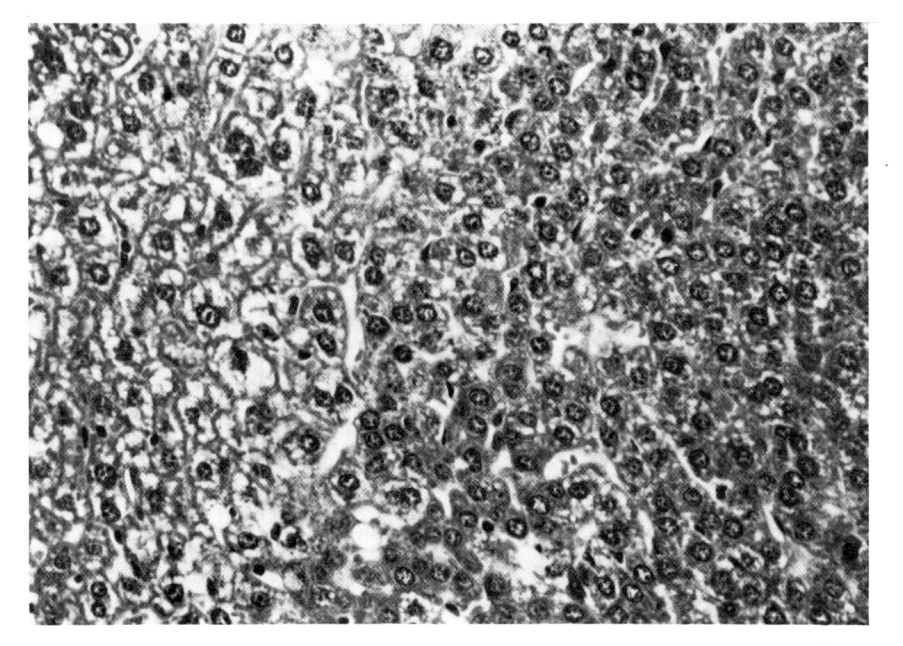

FIGURE 1. Photomicrograph of hepatocellular adenoma in a control mouse composed of small basophilic cells. H&E, X300.

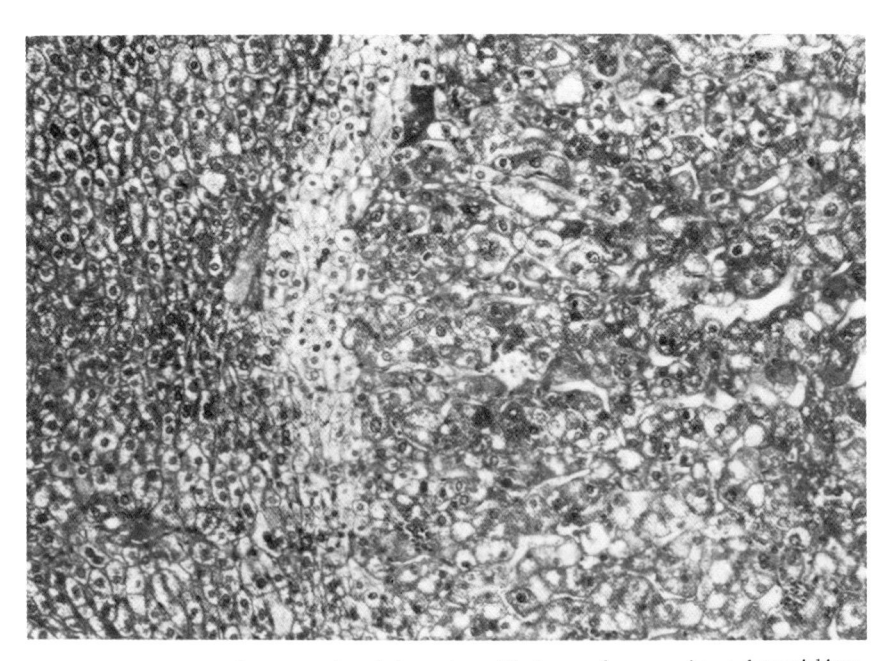

FIGURE 2. Photomicrograph of hepatocellular adenoma in a benzidine-treated mouse composed of large acidophilic cells. H&E, X300.

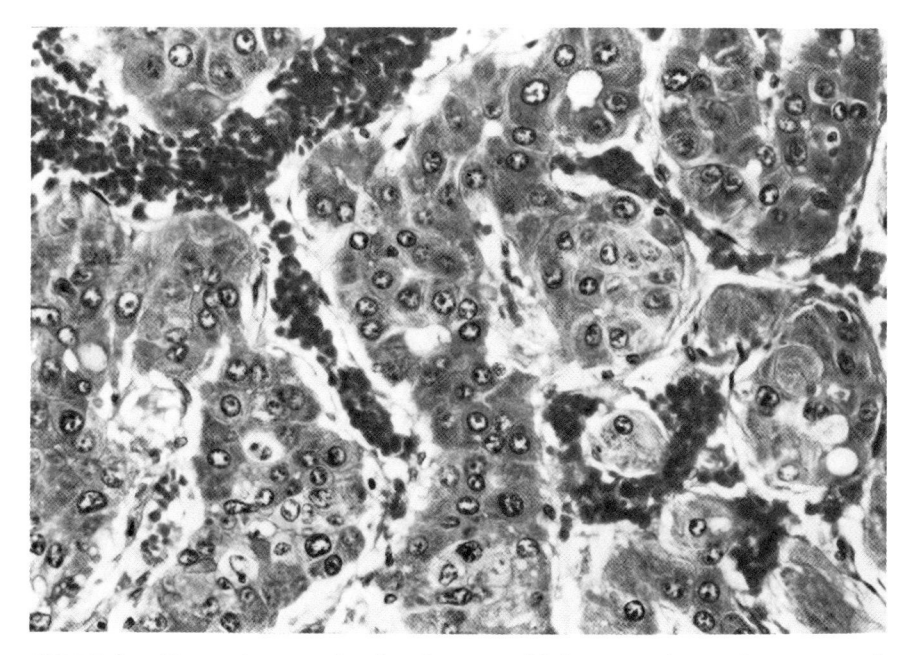

FIGURE 3. Photomicrograph of a hepatocellular carcinoma in a control mouse composed of small basophilic cells. H&E, X300.

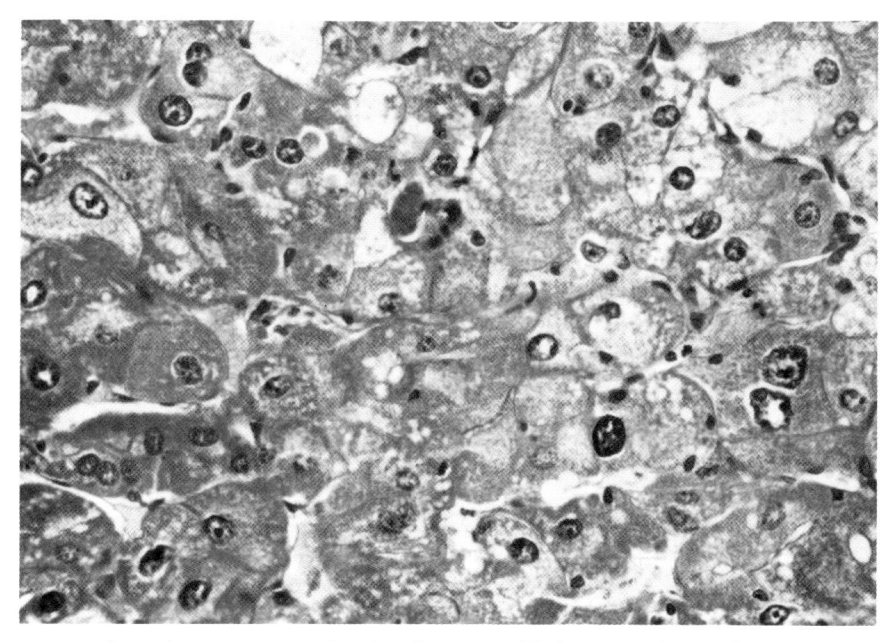

FIGURE 4. Photomicrograph of a hepatocellular carcinoma in a benzidine-treated mouse composed of large acidophilic cells. H&E, X300.

TABLE 2. Comparison of Morphologic Characteristics of Spontaneous and 2-AAF-Induced Hepatocellular Neoplasms in Mice

| Treatment and strain | Number of neoplasms | Ratio of type of characteristics of tumor ||||  Incidence of Trabeculation (%) |
|---|---|---|---|---|---|---|
| | | Basophilic cells | Acidophilic cells | Small cells | Large cells | |
| BALB/c control | 27 | 56:44 | | 89:11 | | 29 |
| BALB/c treated | 45 | 16:84[a] | | 56:44[a] | | 44 |
| $F_1$ Hybrid Control | 11 | 55:45 | | 55:45 | | 22 |
| $F_1$ Hybrid treated | 124 | 18:82[a] | | 32:68 | | 55[a] |

[a] Significantly different from the corresponding control ($p < .05$) utilizing a one-sided test.

The ratio of hepatocellular neoplasms composed of small cells to those of large cells was greater in the control BALB/c mice (89:11) than in the 2-AAF-treated mice (56:44). A similar finding was present in the hepatocellular carcinomas of the control and treated $F_1$ hybrid mice (55:45 compared to 32:68).

A trabecular pattern occurred at an incidence of 29% in the hepatocellular carcinomas in the control BALB/c mice and 44% in the hepatocellular carcinomas in the 2-AAF-treated BALB/c mice (Fig. 5). A similar finding was present in the control and benzidine-treated $F_1$ hybrid mice (22% compared to 55%).

DISCUSSION

A number of investigators have suggested that induced liver tumors in mice are morphologically different from liver tumors in control mice (1-4). It may be of importance to distinguish between a chemical that causes an increased incidence of a naturally occurring tumor from one that induces tumors different from those in the controls. Liver tumors induced with phenobarbital appear to be similar to spontaneous tumors both morphologically and biologically (9). Liver tumors induced with nitrofen (4) appear to be different from naturally occurring hepatocellular carcinomas in the B6C3F1 mouse. In an attempt to help answer this question, a number of morphologic and biologic characteristics associated with liver tumors in control mice and mice treated with 2-AAF or benzidine dihydrochloride were compared.

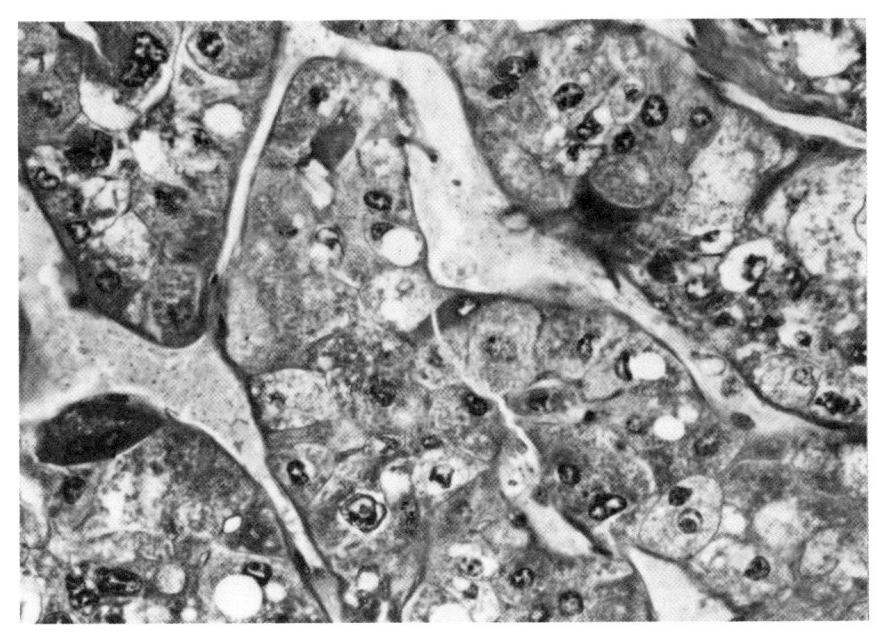

FIGURE 5. Photomicrograph of a prominent trabecular pattern, in a hepatocellular carcinoma, from a benzidine-treated mouse, H&E, X300.

This portion of our study supplied additional information to the previously reported 2-AAF findings (10,11). Tumors induced with either 2-AAF or benzidine tended to be larger and more often multiple than hepatocellular tumors in control mice. The ratio of hepatocellular carcinomas compared to hepatocellular adenomas was slightly increased in the treated mice. Moderately differentiated hepatocellular carcinomas were also more common in the 2-AAF- and benzidine-treated mice, suggesting that the tumors induced with 2-AAF were slightly more malignant than the hepatocellular carcinomas in the control mice.

Liver tumors in the 2-AAF- and benzidine-treated mice were more likely to result in death than those in the controls. This factor may be partially attributable to the fact that the induced tumors were larger and more malignant.

Hoover et al. (4) described consistent morphologic differences between nitrofen-induced and naturally occurring hepatocellular carcinomas in the B6C3F1 mouse. Induced tumors in almost all cases consisted of large eosinophilic hepatocytes containing enlarged and/or hyperchromatic nuclei. In contrast, they reported that hepatocellular carcinomas in control mice consisted of small basophilic cells. In the present study, hepatocellular carcinomas induced with 2-AAF and benzidine also were more often composed of large acidophilic cells. Hepatocellular carcinomas in control mice usually consisted of small basophilic hepatocytes, but some of the hepatocellular carcinomas were composed of acidophilic cells.

Ward et al. (2) also described liver tumors in control mice composed of small basophilic hepatocytes, and hepatocellular carcinomas in mice treated with tetrachlorvinphos composed of solid sheets of large basophilic or eosinophilic hepatocytes. They also described multiple nodules in livers of treated mice compared to single nodules in control mice. Our findings in the present study were similar with regard to size and staining characteristics of the neoplastic hepatocytes. Multiple tumors were also more common than single tumors in the 2-AAF- and benzidine-treated mice than in the controls.

Areas of trabeculation in hepatocellular carcinomas have been associated with an increased incidence of pulmonary metastases (12,13). In this study, trabeculation was more common in the induced tumors than in the naturally occurring tumors. Pulmonary metastases were not seen in the BALB/c mice but did occur at an incidence of 4.8% in the complete study (11). In the $F_1$ hybrid group, the incidence in the controls and benzidine-treated mice was 0 and 7.8%, respectively. Frith et al. (14) also reported that larger liver tumors were more likely to contain prominent areas of trabecular formation and were more likely to metastasize.

In summary, we have identified a number of morphologic and biologic characteristics that were more frequent or less frequent in induced than in naturally occurring hepatocellular tumors. Our results suggest that both 2-AAF and benzidine dihydrochloride may be promoting the incidence of naturally occurring hepatocellular carcinomas, as well as inducing specific hepatocellular carcinomas.

REFERENCES

1. Reznik G, Ward JM: Carcinogenicity of the hair dye component 2-nitro-p-phenylenediamine (2-NPPD): Induction of eosinophilic hepatocellular neoplasms in female B6C3F1 mice. Food Cosmet Toxicol 17:493-500, 1979.

2. Ward JM, Bernal E, Buratto B, Goodman DG, Strandberg JD, Schueler R: Histopathology of neoplastic and nonneoplastic hepatic lesions in mice fed diets containing tetrachlorvinphos. J Natl Cancer Inst 63:111-118, 1979.

3. Reuber MD, Ward JM: Histopathology of liver carcinomas in (C57BL/6N x cC3H/Hen)F1 mice ingesting chlordane. J Natl Cancer Inst 63:89-92, 1979.

4. Hoover KL, Ward JM, Stinson SF: Histopathologic differences between liver tumors in untreated (C57BL/6 x C3H) $F_1$ (B6C3$F_1$) mice and nitrofen-fed mice. J Natl Cancer Inst 65:937-948, 1980.

5. Staffa JA, Mehlman MA (eds.): Innovations in cancer risk assessment. J Environ Pathol Toxicol 3:1-246, 1980.

6. Frith CH, Konvicka AJ: Advances in automation for experimental pathology. Lab Anim Sci 26:171-185, 1976.

7. Frith CH, Herrick S, Konvicka AJ: Computer-assisted collection and analysis of pathology data. J Natl Cancer Inst 58:1717-1727, 1977.

8. Frith CH, Ward JM: A morphologic classification of proliferative and neoplastic hepatic lesions in mice. J Environ Pathol Toxicol 3:329-351, 1980.

9. Peraino C, Fry RJ, Staffeldt T: Enhancement of spontaneous hepatic tumorigenesis in C3H mice by dietary phenobarbital. J Natl Cancer Inst 51:1349-1350, 1973.

10. Littlefield NA, Farmer JH, Gaylor DW, Sheldon WG: Effects of dose and time in a long-term, low-dose carcinogenic study. J Environ Pathol Toxicol 3:17-34, 1980.

11. Frith CH, Kodell RL, Littlefield NA: Biologic and morphologic characteristics of hepatocellular lesions in BALB/c mice fed 2-acetylaminofluorene. J Environ Pathol Toxicol 3:121-138, 1980.

12. Vesselinovitch SD, Mihailovich N, Rao KVN: Morphology and metastatic nature of induced hepatic nodular lesions in C57BL x C3H F1 mice. Cancer Res 38:2003-2010, 1978.

13. Ward JM, Vlahakis G: Evaluation of hepatocellular neoplasms in mice. J Natl Cancer Inst 61:807-811, 1978.

14. Frith CH, Baetcke KP, Nelson CJ, Schieferstein G: Correlation of liver tumor morphology and weight to incidence of pulmonary metastases in the mouse. Toxicol Lett 7:113-118, 1980.

Chapter 7

# DNA Alkylation and Cell Replication in Mouse Liver Carcinogenesis

**James A. Swenberg
and Charles Lindamood III**

INTRODUCTION

The widespread use of mice in carcinogenicity studies, coupled with the frequent occurrence of hepatic neoplasms, has brought about a critical need to better understand the mechanisms involved in the induction of mouse liver cancer. Recent studies have demonstrated that rodent hepatocarcinogenesis is a multistage process involving initiation and promotion (1-3). By definition, initiation represents a permanent heritable change that is thought to involve mutational events. In contrast, promotion is thought to be a reversible process whereby initiated cells undergo selective growth. Cell replication appears to be critical for both initiation and promotion. This paper will review the present state of knowledge regarding DNA damage, repair, and replication in mouse liver following exposure to alkylating agents.

DNA ALKYLATION AND REPAIR

Alkylation occurs at multiple sites on the DNA (4-9). The proportion of nitrogen to oxygen alkylation depends on the chemical's reactivity, with weak $S_N 2$ agents reacting almost exclusively with strong nucleophilic sites such as the N-7 position of guanine. In contrast, strong $S_N 1$-type methylating agents alkylate nitrogen and oxygen atoms of DNA. Relatively greater alkylation of oxygen takes place with ethylating agents, although total alkylation is frequently less. The site of DNA alkylation appears to be critical, since alkylation at several positions has been shown to result in mispairing using RNA or DNA polymerase systems (5,9-11), while alkylation at other sites does not. Lesions on the DNA that result in mispairing when DNA replication occurs prior to their repair are termed promutagenic lesions. $O^6$-Alkylguanine ($O^6$AG) is the most widely studied promutagenic lesion produced by alkylating agents. There is a good correlation between susceptibility of a given organ to tumor formation following exposure to alkylating agents and persistence and/or accumulation of $O^6$AG (5,6,12-14). Furthermore, mutagenesis by N-methyl-N'-nitro-N-nitrosoguanidine in E. coli was correlated with the inability to repair $O^6$-methylguanine ($O^6$MG) (15), with mutations being primarily due to GC → AT transitions (16).

While the correlation between persistence of $O^6$AG and target organ for carcinogenicity has been high, a prominent exception has been the rat liver. $O^6$AG is removed from rat liver more efficiently than from any other tissue, even though many alkylating agents induce hepatic tumors (5,13,14). Mouse liver responds similarly; however, it has not been investigated in as much detail. $O^6$MG was removed rapidly from mouse liver following exposure to methylnitrosourea (MNU) at 10-25 mg/kg but slowly when 80 mg MNU/kg were administered (17,18). These data are similar to those obtained with rats (19-22) and suggest that the DNA repair system for $O^6$MG is saturable. Recent investigations have demonstrated that both rat (20-26) and mouse (27) liver contain an alkyl acceptor protein that stoichiometrically removes the alkyl group from $O^6$AG, leaving the parent nucleotide in the DNA. The alkyl group is bound to a cysteine residue of the protein. At high concentrations of $O^6$AG, the alkyl acceptor protein is consumed much more rapidly than it is formed, leading to persistence of $O^6$AG in liver DNA. At lower concentrations, the alkyl group is rapidly removed from $O^6$AG.

Several other promutagenic lesions are only slowly removed from rodent liver. $O^2$-Ethylthymine and $O^4$-ethylthymine are reported to have half-lives in rat liver of 48 hours (28) to 19 days (29). Only $O^2$-ethylcytosine has been investigated in mouse liver, where it persisted for 196 hours (18).

CELL REPLICATION IN INITIATION AND PROMOTION

The presence of promutagenic DNA damage is not, however, enough to cause initiation. Rather, it is the amount of promutagenic damage present during cell replication - more specifically, during de novo DNA synthesis - that determines the probability of a mutation occurring. Of crucial importance to this theory is the demonstration that alkylated DNA can be replicated. This has been demonstrated in several in vitro systems (5,9,11), as well as in rat liver (30). Further support for the requirement of cell replication comes from the experiments of Craddock, who demonstrated that liver tumors could only be induced by a single dose of nitroso compounds if the chemical was administered in conjunction with a partial hepatectomy (31). Similarly, cell proliferation was a necessary requirement for the induction of hepatic foci by MNU and 1,2-dimethylhydrazine (SDMH) (32). The compensatory restorative hyperplasia that is a sequela to the necrogenic effects of nitrosamines may be a factor in the production of preneoplastic foci (33) and carcinogenicity in the liver (34). Differential sensitivity of two strains of mice to lung and liver neoplasia following DMN administration has also been correlated with increased de novo DNA synthesis in the strain's target organ (35).

Increased cell replication can also be a promoting factor, as recently demonstrated for skin (36). Several chemicals that are devoid of genotoxic activity do result in increased tumor incidences when chronically administered to rodents. These chemicals have been termed epigenetic carcinogens, but may actually be promotors acting on background or spontaneously initiated cells. Whether or not they selectively enhance replication of initiated cells is unknown.

CELL SPECIFICITY IN HEPATOCARCINOGENESIS

When looking for biochemical mechanisms responsible for hepatocarcinogenesis, it is important to remember that hepatocytes vastly outnumber the other cell types of liver. Hepatocytes account for nearly 90% of the liver's mass and 60-70% of its cells (37). The remaining nonparenchymal cells consist of endothelial, Kupffer, and bile duct cells. The endothelial cells are present in a 2:1 ratio with Kupffer cells and together contain 10-20% of the liver's DNA, while bile duct cells comprise only about 1%. Since many hepatocarcinogens are relatively specific for inducing hepatocellular carcinomas versus angiosarcomas, it is necessary to examine DNA alkylation, repair and replication in specific cell populations. Several methods are available to separate hepatocytes and nonparenchymal cells (NPC) following collagenase perfusion of the liver. Hepatocytes can be selectively digested by pronase, yielding large numbers of NPC (38). The mixed liver-cell suspension obtained by collagenase perfusion can be separated into hepatocytes and NPC by centrifugal elutriation (39,40) or by differential centrifugation (41,42). The NPC obtained by the latter method can be further separated into endothelial and Kupffer cells by elutriation (39). Thus, it is possible to critically evaluate biochemical mechanisms of chemical carcinogenesis by comparing endpoints in target and nontarget cells within the target organ.

Recent studies from our laboratory on rats continuously exposed to SDMH have demonstrated that $O^6MG$ selectively accumulates in the NPC, but not in hepatocytes (43) and that a pronounced mitogenic response occurs in NPC during SDMH exposure (44). Thus, it is highly probable that mutations arising from replication of DNA containing $O^6MG$ represent a major factor in angiosarcoma induction by SDMH. On the other hand, exposure of rats to 40 ppm DEN in their drinking water did not result in accumulation of $O^6$-ethylguanine ($O^6EG$) in hepatocytes or NPC, suggesting that other promutagenic lesions may play the major role in initiating hepatocytes.

We have used similar techniques to evaluate cell-specific differences in dimethylnitrosamine- (DMN) induced hepatocarcinogenesis in C3H mice. Male and female C3H/HeNCrlBR mice were exposed to 10 ppm DMN in their drinking water for up to 32 days, during which $O^6MG$ concentrations and de novo DNA synthesis were followed (42). This dosing regimen induced hemangiomas and angiosarcomas in the livers of male and female mice (45). Males, but not females, were reported to also develop hepatocellular carcinomas; however, a second report from the same laboratory failed to confirm this (46). Interpretation of the hepatocellular response in male C3H mice is complicated by their high spontaneous incidence of hepatocellular carcinoma (47). It is not known whether the enhancement of hepatocellular carcinoma in the earlier study was a direct effect of DMN or an exacerbation of the spontaneous incidence.

After 2, 4, 8, 16, or 32 day exposures to 10 ppm DMN, the mice received one intraperitoneal injection of [$^3$H]thymidine (2.5 µCi/g) 1 hour before sacrifice (42). Briefly, their livers were perfused with collagenase and the mixed liver-cell suspension was separated into NPC and hepatocytes using differential centrifugation (70 x g, 2 min). DNA was isolated from the cells using hydroxyapatite chromatography. Estimations of de novo DNA synthesis were accomplished by determining the radioactivity per µg DNA, while O$^6$MG and 7-methylguanine were detected by fluorescence spectrometry after applying 0.1 N HCl hydrolysates of DNA to two strong cation exchange HPLC columns in series.

A typical chromatogram illustrating the separation of normal and alkylated purines is shown in Fig. 1. All treatment groups had

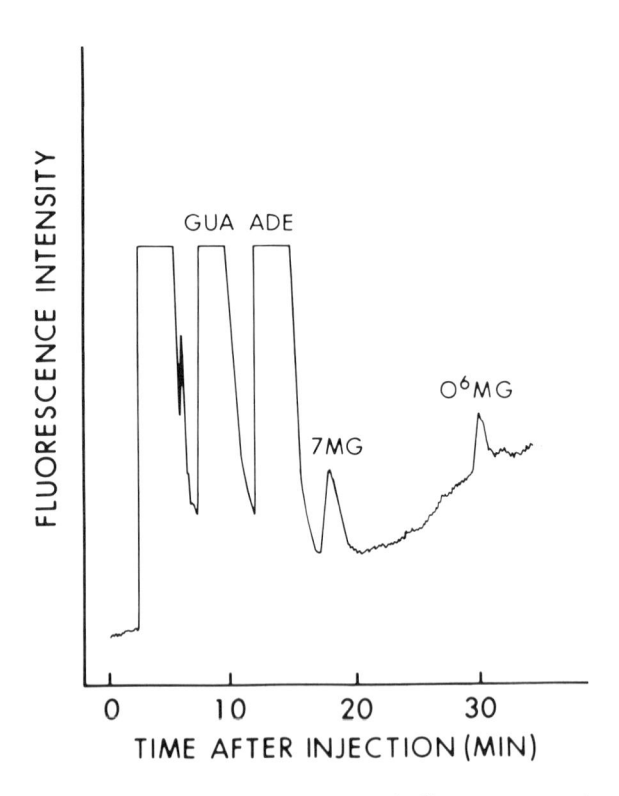

FIGURE 1. HPLC separation and fluorescence detection of guanine (Gua), adenine (Ade), 7-methylguanine (7MG), and O$^6$-methylguanine from hepatocytes of mice exposed to 10 ppm DMN for 16 days.

7-methylguanine (7MG) and $O^6MG$ concentrations above the limits of detectability. 7MG values were not available for all samples, due to coelution of an unknown peak with 7MG. Regression analysis of the alkylation experiments indicated that sex differences were not significant. Hence, combined data from male and female mice are shown in Table 1. 7MG concentrations increased from 97 pmol/mg DNA at 2 days to 158 pmol/mg DNA at 16 days in the NPC. The hepatocytes were similar where 7MG increased from 86 pmol/mg DNA at 2 days to 221 pmol/mg DNA at 8 days. By 32 days, hepatocyte 7MG concentrations had declined slightly to 183 pmol/mg DNA. Although hepatocyte concentrations of 7MG were slightly higher than NPC, no significant differences in 7MG concentrations were demonstrated between cell types.

In contrast, there were significant differences in $O^6MG$ concentrations between the hepatocytes and NPC. Concentrations of $O^6MG$ in the NPC increased from 13 pmol/mg DNA at 2 days to 51 pmol/mg DNA at 32 days (Table 1), while concentrations of $O^6MG$ in hepatocytes remained relatively constant over the course of DMN exposure. Values ranged from 9 pmol/mg DNA at 4 days to 4 pmol/mg DNA at 32 days.

Because the concentrations of 7MG were not statistically different between the two liver cell types, calculation of ratios of $O^6MG$ to 7MG revealed distinct differences in the ability of hepatocytes and NPC to remove $O^6MG$ (Table 1). $O^6MG/7MG$ ratios increased from 0.13 at 2 d to 0.27 at 16 days in the NPC. On the other hand, the $O^6MG/7MG$ ratios in hepatocytes declined slightly from 0.06 at 2 days to 0.02 at 32 days. The hepatocyte ratios were all lower than 0.1, indicating preferential removal of $O^6MG$. In contrast, ratios of $O^6MG/7MG$ for NPC were considerably higher than 0.1 and 4 to 10 times higher than corresponding hepatocytes. This demonstrates that the NPC population is less able to remove $O^6MG$ than hepatocytes.

No sex differences in de novo DNA synthesis were demonstrable, as a result of DMN exposure, in C3H mice. Significant differences in de novo DNA synthesis occurred between NPC and hepatocytes as a result of DMN exposure. De novo DNA synthesis in hepatocytes increased at 4 days (141 versus 40 dpm/μg DNA for control hepatocytes) and remained slightly elevated at 32 days (98 dpm/μg DNA) of DMN exposure. De novo DNA synthesis increased progressively between 4 (177 dpm/μg DNA) and 32 days (400 dpm/μg DNA) in the NPC.

The initiation index, defined as the product of $O^6MG$ concentration, de novo DNA synthesis, and the amount of DNA at risk, has been calculated for NPC and hepatocytes (42). During the first 16 days of DMN exposure, NPC and hepatocytes had approximately equal probabilities of becoming initiated. As exposure to DMN continued, the probability of initiation in NPC relative to hepatocytes increased such that the initiation index was more than twice as high in NPC than in hepatocytes after 32 days of DMN exposure. This is consistent with the results of bioassays in which males and females of a variety of mouse strains developed hemangiomas and angiosarcomas in response to DMN exposure, while the development of hepatocellular carcinoma was both strain- and dose-dependent (47) and varied in the same laboratory from year to year (45,46).

TABLE 1. Alkylation and De Novo Synthesis of Liver Cell DNA from C3H Mice During Continuous DMN Exposure

| Cell type | Days of exposure | pmol 7MG/mg DNA[a] | pmol O$^6$MG/mg DNA[b] | O$^6$MG/7MG | De novo DNA synthesis (dpm/μg DNA)[c] |
|---|---|---|---|---|---|
| NPC | 0 | – | – | – | 123 + 6 |
|  | 2 | 97 + 26 | 13 + 1 | 0.13 | 109 + 11 |
|  | 4 | 134 + 44 | 26 + 3 | 0.19 | 177 + 35 |
|  | 8 | N.D. | 27 + 3 | N.D. | 227 + 23 |
|  | 16 | 158 + 43 | 43 + 3 | 0.27 | 260 + 25 |
|  | 32 | N.D. | 51 + 2 | N.D. | 400 + 27 |
| Hepatocyte | 0 | – | – | – | 40 + 4 |
|  | 2 | 86 + 8 | 5 + 1 | 0.06 | 58 + 7 |
|  | 4 | 152 + 23 | 9 + 2 | 0.06 | 141 + 37 |
|  | 8 | 221 + 21 | 6 + 1 | 0.03 | 128 + 25 |
|  | 16 | 214 + 41 | 5 + 1 | 0.02 | 128 + 42 |
|  | 32 | 183 + 20 | 4 + 1 | 0.02 | 98 + 17 |

[a] Mean + SEM for 2–8 determinations.
[b] Mean + SEM for 4–8 determinations.
[c] Mean + SEM for 5–13 animals.
N.D. = not determined.

Comparisons of this mouse data and a previous rat study (43) suggested that there were differences in the degree of inducibility of enzymatic repair of $O^6MG$ in mouse and rat hepatocytes as a consequence of exposure to methylating agents. Exposure of rats to 30 ppm 1,2-dimethylhydrazine in the drinking water induced rapid and extensive repair of $O^6MG$ in hepatocytes, resulting in progressively lower concentrations of $O^6MG$. In contrast, $O^6MG$ concentrations remained virtually constant in mouse hepatocellular DNA during exposure to 10 ppm DMN. The enhancement of $O^6AG$ alkyl acceptor protein activity in rat hepatocytes was confirmed by direct assay of the enzyme, where a 2-3-fold increase in $O^6AG$ alkyl acceptor protein activity occurred during continuous 1,2-dimethylhydrazine exposure (26). We have subsequently demonstrated that $O^6AG$ alkyl acceptor protein activity in C3H mouse hepatocytes is not enhanced during continuous exposure to 10 ppm DMN (Fig. 2). The lack of $O^6AG$ alkyl acceptor protein enhancement in mouse hepatocytes during periods of carcinogen induced <u>de novo</u> DNA synthesis demonstrates that enhancement of $O^6MG$ repair is not a secondary response to increased cell proliferation in mouse liver (42).

Differences in the enzymatic repair of $O^6MG$ may represent important molecular events when comparing species differences in susceptability to carcinogenic insult. Mouse hepatocytes are less able to repair $O^6MG$ than rat hepatocytes, which in turn have only one-tenth the ability of human liver (48) to repair $O^6MG$. Thus, exposures to

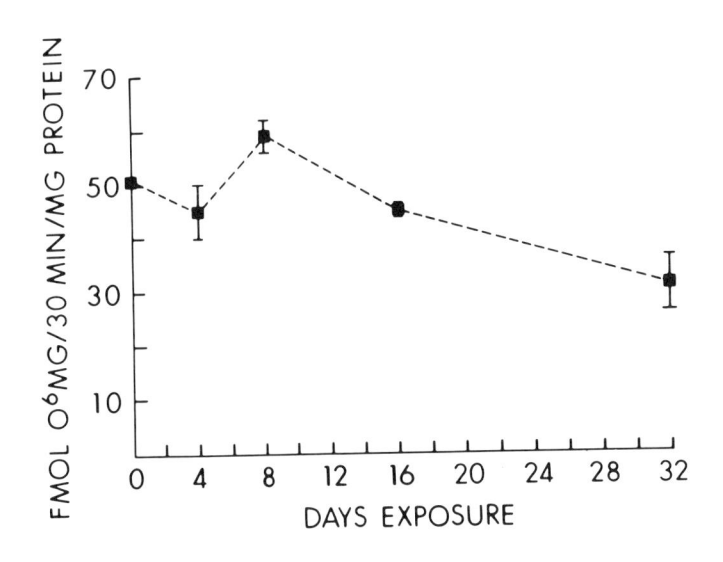

FIGURE 2. $O^6MG$ alkyl acceptor protein activity in hepatocytes from C3H mice exposed to 10 DMN for up to 32 days. Each point is the mean ± SEM for 3 animals.

methylating agents, resulting in similar concentrations of $O^6MG$ in mouse and human hepatic DNA, would be expected to initiate many more mouse hepatocytes than human hepatocytes. Since alkylating agents produce several other promutagenic lesions, similar comparative data on the repair of these will be needed before such data can be used for quantitative risk assessment.

REFERENCES

1. Farber E: The sequential analysis of liver cancer induction. Biochim Biophys Acta 605:149-166, 1980.

2. Pitot HC, Sirica AE: The stages of initiation and promotion in hepatocarcinogenesis. Biochim Biophys Acta 605:191-215, 1980.

3. Popp JA, Leonard TB: Potential use of initiation-promotion studies in understanding mouse liver neoplasia. Chap. 8 in this volume.

4. Margison GP, O'Connor PJ: Nucleic acid modification by N-nitroso compounds. In: Chemical Carcinogens and DNA, edited by PL Grover, vol. 1, pp. 111-159. Boca Raton, Fla.: CRC Press, 1979.

5. Montesano R: Alkylation of DNA and tissue specificity in nitrosamine carcinogenesis. J Supramol Struct Cell Biochem 17:259-273, 1981.

6. Pegg AE: Formation and metabolism of alkylated nucleosides: Possible role in carcinogenesis by nitroso compounds and alkylating agents. Adv Cancer Res 25:195-269, 1977.

7. Singer B: The chemical effects of nucleic acid alkylation and their relation to mutagenesis and carcinogenesis. In: Progress in Nucleic Acid Research and Molecular Biology, vol. 15, pp. 219-284. New York: Academic, 1975.

8. Singer B: Sites in nucleic acids reacting with alkylating agents of differing carcinogenicity or mutagenicity. J Toxicol Environ Health 2:1279-1295, 1977.

9. Singer B, Kusmierek JT: Chemical mutagenesis. Ann Rev Biochem 52:655-693, 1982.

10. Gerchman LL, Ludlum DB: The properties of $O^6$-methylguanine in templates for RNA polymerase. Biochim Biophys Acta 308:210-216, 1973.

11. Abbott PJ, Saffhill R: DNA synthesis with methylated poly(dC-dG) templates. Evidence for a competitive nature to miscoding by O6-methylguanine. Biochem Biophys Acta 562:51-61, 1979.

12. Kleihues P, Doerjer G, Swenberg JA, Hauenstein E, Bucheler J, Cooper HK: DNA repair as regulatory factor in organotrophy of alkylating carcinogens. Arch Toxicol Suppl 2:253-261, 1979.

13. Margison GP, Saffhill R: Carcinogenicity of alkylating agents. In: Advances in Medical Oncology, Research and Education, vol. 1, Carcinogenesis, edited by GP Margison, pp. 229-239. New York: Pergamon, 1978.

14. Pegg AE, Nicoll JW: Nitrosamine carcinogenesis: The importance of the persistence in DNA of alkylated bases in the organotropism of tumour induction. In: Screening Tests in Chemical Carcinogenesis, IARC Scientific Publication No. 12, pp. 571-590. Lyon: IARC, 1976.

15. Schendel PG, Robins PE: Repair of $O^6$-methylguanine in adapted Escherichia coli. Proc Natl Acad Sci 75:6017-6020, 1978.

16. Coulondre C, Miller JH: Genetic studies of the lac repressor IV. Mutagenic specificity in the lacI gene of Escherichia coli. J Mol Biol 117:577-606, 1977.

17. Bücheler J, Kleihues P: Excision of $O^6$-methylguanine from DNA of various mouse tissues following a single injection of N-methyl-N-nitrosourea. Chem-Biol Interact 16:325-333, 1977.

18. Frei JV, Swenson DH, Warren W, Lawley PD: Alkylation of deoxyribonucleic acid in vivo in various organs of C57Bl mice by the carcinogens N-methyl-N-nitrosourea, N-ethyl-N-nitrosourea, and ethyl methanesulphonate in relation to induction of thymic lymphoma. Biochem J 174:1031-1044, 1978.

19. Kleihues P, Margison GP: Exhaustion and recovery of repair excision of $O^6$-methylguanine from rat liver DNA. Nature (Lond) 259:153-155, 1976.

20. Pegg AE: Enzymatic removal of $O^6$-methylguanine from DNA by mammalian cell extracts. Biochem Biophys Res Commun 84:166-173, 1978.

21. Pegg AE: Dimethylnitrosamine inhibits enzymatic removal of $O^6$-methylguanine from DNA. Nature (Lond) 274: 182-184, 1978.

22. Pegg AE, Balog B: Formation and subsequent excision of $O^6$-ethylguanine from DNA of rat liver following administration of diethylnitrosamine. Cancer Res 39:5003-5009, 1979.

23. Pegg AE, Hui G: Formation and subsequent removal of $O^6$-methylguanine from deoxyribonucleic acid in rat liver and kidney after small doses of dimethylnitrosamine. Biochem J 173:739-748, 1978.

24. Pegg AE, Perry W: Stimulation of transfer of methyl groups from $O^6$-methylguanine in DNA to protein by rat liver extracts

in response to hepatotoxins. Carcinogenesis 2:1195-1200, 1981.

25. Pegg AE, Perry W, Bennett RA: Effect of partial hepatectomy on removal of $O^6$-methylguanine from alkylated DNA by rat liver extracts. Biochem J 197:195-201, 1981.

26. Swenberg JA, Bedell MB, Billings KC, Umbenhauer DR, Pegg AE: Cell specific differences in $O^6$-alkylguanine DNA repair activity during continuous carcinogen exposure. Proc Natl Acad Sci 79:5499-5502, 1982.

27. Bogden JB, Eastman A, Bresnik E: A novel system in mouse liver for the repair of $O^6$-methylguanine lesions in methylated DNA. Nucl Acid Res 9:3089-3103, 1981.

28. Singer B, Spengler S, Bodell WJ: Tissue-dependent enzyme-mediated repair of removal of O-ethylpyrimidines and ethylpurines in carcinogen-treated rats. Carcinogenesis 2:1069-1073, 1981.

29. Scherer E, Timmer AP, Emmelot P: Formation by diethylnitrosamine and persistence of $O^4$-ethylthymidine in rat liver DNA in vivo. Cancer Lett 10:1-6, 1980.

30. Abanobi S, Farber E, Sarma DSR: In vitro replication of hepatic deoxyribonucleic acid of rats treated with dimethylnitrosamine: Presence of dimethylnitrosamine-induced $O^6$-methylguanine, $N^7$-methylguanine and $N^3$-methylguanine in the replicated hybrid deoxyribonucleic acid. Cancer Res 39:1592-1596, 1979.

31. Craddock VM: Induction of liver tumours in rats by a single treatment with nitroso compounds given after partial hepatectomy. Nature (Lond) 245:386-388, 1973.

32. Columbano A, Rajalakshmi S, Sarma DSR: Requirement of cell proliferation for the initiation of liver carcinogenesis as assayed by three different procedures. Cancer Res 41:2079-2083, 1981.

33. Ying TS, Sarma DSR, Farber E: Role of acute hepatic necrosis in the induction of early steps in liver carcinogenesis by diethylnitrosamine. Cancer Res 41:2096-2102, 1981.

34. Craddock VM, Henderson AR: De novo replication and repair of DNA during diethylnitrosamine induced carcinogenesis. Cancer Lett 3:277-284, 1977.

35. De Munter HK, Den Engelse L, Emmelot P: Studies on lung tumours. IV. Correlation between [$^3$H]-thymidine labelling of lung and liver cells and tumour formation in GRS/A and C3Hf/A mice following administration of dimethylnitrosamine. Chem-Biol Interact 24:299-316, 1979.

36. Argyris TS, Slaga TJ: Promotion of carcinomas by repeated abrasion in initiated skin of mice. Cancer Res 41:5193-5195, 1981.

37. Wisse E, Knook DL: The investigation of sinusoidal cells, a new approach to the study of liver function. In: Progress in Liver Disease, vol. 6, edited by H Popper, F Schaffner, pp. 153-171. New York: Grune & Stratton, 1979.

38. Mills DM, Zucker-Franklin D: Electron microscopic study of isolated Kupffer cells. Am J Pathol 54:147-155, 1969.

39. Knook DL, Sleyster E: Separation of Kupffer and endothelial cells by centrifugal elutriation. Exp Cell Res 99:444-449, 1976.

40. Lewis JG, Swenberg JA: Differential repair of $O^6$-methylguanine in DNA of rat hepatocytes and nonparenchymal cells. Nature (Lond) 288:185-187, 1980.

41. Knook DL, Sleyster EC: Preparation and characterization of Kupffer cells from rat and mouse liver. In: Kupffer Cells and Other Liver Sinusoidal Cells, edited by E Wisse, DL Knook, pp. 273-288. Amsterdam: Elsevier/North-Holland Biomedical Press, 1977.

42. Lindamood C, Bedell M, Billings KC, Swenberg JA: Alkylation and de novo synthesis of liver cell DNA from C3H mice during chronic dimethylnitrosamine exposure. Cancer Res 42:4153-4157, 1982.

43. Bedell MA, Lewis JG, Billings KC, Swenberg JA: Cell specific hepatocarcinogenesis: $O^6$-Methylguanine preferentially accumulates in target cell DNA during continuous exposure of rats to 1,2-dimethylhydrazine. Cancer Res 42:3079-3083, 1982.

44. Lewis JG, Swenberg JA: Effect of 1,2-dimethylhydrazine and diethylnitrosamine on cell replication and unscheduled DNA synthesis in target and nontarget cell populations in rat liver following chronic administration. Cancer Res 42:89-92, 1982.

45. Den Engelse L, Hollander CF, Misdorp W: A sex-dependent difference in the type of tumours induced by dimethylnitrosamine in the livers of C3Hf mice. Eur J Cancer 10:129-135, 1974.

46. Den Engelse L, Oomen, LCJM, Van Der Valk MA, Hart AAM, Dux A, Emmelot P: Studies on lung tumours. V. Susceptibility of mice to dimethylnitrosamine-induced tumour formation in relation to H-2 haplotype. Int J Cancer 28:199-208, 1981.

47. Grasso P, Hardy J: Strain difference in natural incidence of response to carcinogens. In: Mouse Hepatic Neoplasia, edited by W.H. Butler, pp. 111-132. Amsterdam: Elsevier, 1975.

48. Pegg AE, Roberfroid M, von Bahr C, Foote RS, Mitra S, Bresil H, Likhachev A, Montesano R: Removal of $O^6$-methylguanine from DNA by human liver fractions. Proc Natl Acad Sci 79:5162-5165, 1982.

# Chapter 8

# Potential Use of Initiation-Promotion Studies in Understanding Mouse Liver Neoplasia

## James A. Popp and Thomas B. Leonard

The concept of initiation and promotion in neoplastic formation has been well established during the last forty years of carcinogenesis research. The terms "initiate" and "promote" were apparently first applied to the study of neoplasm formation around sites of tissue damage on rabbit ears previously treated with topical application of coal tars (1,2). However, it was the subsequent studies of Berenblum and Shubik (3,4) that clearly defined the two stages in neoplastic development. In ensuing studies, the two stages have been further characterized and the properties of initiating and promoting agents have been further defined (5). Initiation is an irreversible cellular alteration, probably mediated through changes in DNA. The "irreversible alteration" is necessary for subsequent neoplastic formation to occur, but is not sufficient to induce a neoplasm without some further stimulus. In contrast, promotion is a reversible process in which a neoplastic lesion develops from a previously altered but probably inapparent initiated cell.

The classical definitions state that neither a pure initiator nor a pure promoter is capable of inducing a neoplasm by itself, even though the steps of initiation and promotion are necessary for most, if not all, neoplastic development. At the operational level, several other characteristics of initiation and promotion are apparent. The initiation phase must occur before the promoting agent can have any effect. Since the initiated cell does not revert to normal, the promoting stimulus may be applied at variable time intervals after initiation and result in neoplastic formation. Since the promotional event is reversible, multiple exposures of the promotional agent are required to affect promotion. Although the stages of initiation and promotion are separable in many experimental protocols using specified and well-defined agents, it should be remembered that a neoplasm in a target organ formed after prolonged exposure to a complete carcinogen is the result of the initiating and promoting activity derived from the single chemical agent.

Multiple steps have been postulated and discussed for neoplastic development in the rodent liver for a number of years (6). However, initiation-promotion studies designed to analyze the steps involved in hepatocarcinogenesis began only after the phenobarbital enhancement of 2-acetylaminofluorene- (2-AAF) initiated lesions was demonstrated in the rat (7). In these studies, a 2-week feeding of 2-AAF was used as the initiating regimen, which alone resulted in a low number of hepatic neoplasms. However, if phenobarbital feeding followed the 2-AAF treatment, a much higher incidence of hepatocellular neoplasms was observed, while feeding phenobarbital alone failed to induce any neoplasms. Subsequent studies (8,9) indicated that phenobarbital was acting as a promoting agent and was probably devoid of any initiating activity. Pitot et al. (10) have more recently adopted the use of phenobarbital as a promoting agent in a system where a single administration of diethylnitrosamine (DEN) to partially hepatectomized rats forms the initiating regimen. This system has been used extensively to characterize the sequential development of rat neoplastic lesions. Various modifications of the systems originally used by Peraino and Pitot have identified promoting activity, or at least neoplastic enhancing activity, in several different chemical agents (9,11).

Two other systems have been described in which promotion of initiated hepatocytes to neoplastic lesions occurs. Solt and Farber (12) described a system in which a necrogenic dose of DEN was used as the initiating agent. After a 2-week recovery period, the rats were placed on a 2-AAF-containing diet for 2 weeks. A partial hepatectomy was performed 1 week after 2-AAF feeding began. Since AAF suppresses proliferation of normal but not altered (initiated) hepatocytes, the altered cells rapidly proliferate to form large foci and small nodules within 1 month following the initiating dose of DEN. Although the second phase of this system (2-AAF plus partial hepatotectomy) is usually referred to as a "growth selection" process, it clearly has a neoplastic enhancing effect (13). Modification of this system as described by Cayama et al. (14) allows the use of this system to study neoplastic development with nonhepatonecrogenic agents. In 1979, Shinozuka et al. (15) described the enhancement of hepatic neoplastic formation by the feeding of choline-devoid diet. The hepatic mitogenic effect of choline-devoid diet compared to the lack of a mitogenic effect of phenobarbital suggest that different mechanisms of promotion may exist for these two promotion systems (16).

Although several chemicals have been described as promoters of hepatocarcinogenesis, a semantic problem arises for some of these chemicals, as well as future chemicals which may be tested for promotional activity. A promoter, by definition, does not cause neoplasm formation unless initiated cells exist to receive the affect of the promoting agent. In other words, the promoter is not a complete carcinogen, and the term "promoter" should not be applied to express the promoting activity of a complete carcinogen. However, use of the initiation-promotion systems to define the promotional activity of a complete carcinogen is useful. To avoid semantic confusion, an agent demonstrating promoting activity in an initiation-promotion regimen in the liver could be referred to as an "enhancing agent" or as a chemical containing "promoting activity" until it is clear whether the agent is a true promoter or a carcinogen. The term "promoter" should be reserved for those agents that have significant promoting activity yet fail to demonstrate initiating activity or act as a complete carcinogen.

Although not immediately useful as a screening test, the judicious use of initiation-promotion systems can provide information on the mechanism of action of a known or suspected hepatocarcinogenic agent. The initiation-promotion systems also have the potential to define the degree of initiating and promoting activity contained in a known or suspected hepatocarcinogenic agent. As described by Farber (17) and by Pitot and Sirica (9), these systems provide a method to separate the stages of neoplastic development. This allows the study of various stages and advances our understanding of the progressive development of neoplastic lesions. By using a known promoting agent on a population of preneoplastic lesions, modulation of the lesion either forward to a neoplasm or backward to "normal liver" can be studied to define the biological potential of the various preneoplastic lesions.

Despite the advantages of initiation-promotion systems in defining the biology of preneoplastic lesions and the mechanism of action of hepatocarcinogens as demonstrated in the rat liver systems, little published information is available concerning initiation and promotion in mouse liver neoplasm development. Phenobarbital has tumor enhancing activity in both sexes of C3H mice, which develop spontaneous hepatic neoplasms (Table 1) (18). Phenobarbital at 0.05% in the diet accelerates the expression of hepatic tumorigenesis (18,19) and also causes the appearance of gamma-glutamyl transpeptidase-positive (GGT+) foci in many of the neoplasms. The spontaneous neoplasms appearing in this strain of mouse uniformly lack GGT+ staining. In two additional studies, phenobarbital was given to CF-1 mice via either the water (20) or the diet (21). Again, phenobarbital was shown to enhance spontaneous hepatic tumorigenesis, with hepatic neoplasms appearing earlier and at a higher cumulative incidence in phenobarbital-treated mice. The average age of death with hepatomas was 84.9 weeks for phenobarbital-treated male mice and 105 weeks for control male mice. The promotional effect of phenobarbital was organ-specific, since the number of hepatomas was increased while the numbers of other spontaneous neoplasms were unaltered (Table 2). These four studies have focused on the tumor-enhancing property of phenobarbital on spontaneous hepatic neoplasms rather than on chemically initiated neoplasms.

In a single reported study, the promotional effect of phenobarbital has been determined in chemically-initiated sites in the mouse liver (22). Inbred DDD newborn mice were injected with 3 µl of dimethylnitrosamine. Starting at weaning, the mice were given 0.05% phenobarbital in the drinking water. Mice were killed at 16 weeks of age. Phenobarbital enhanced the number of liver neoplasms (Table 3) but not lung neoplasms. No sex differences were noted in neoplasms either with or without phenobarbital. These results demonstrate that phenobarbital may act as a neoplastic enhancing

TABLE 1. Hepatic Tumor Incidence[a]

| Phenobarbital in diet (%) | Sex | Number of Mice |||  Percent with tumors | Number of tumors ||
|---|---|---|---|---|---|---|---|
| | | Per cage | Per group | With tumors | | Total/group | Average/mouse |
| 0 | F | 5 | 16 | 1 | 6 | 1 | 0.06 |
| 0.05 | F | 5 | 16 | 10 | 63 | 23 | 1.44 |
| 0 | M | 5 | 17 | 7 | 41 | 7 | 0.41 |
| 0.05 | M | 5 | 17 | 16 | 94 | 76 | 4.47 |

[a]Adapted from Peraino et al. (18).

TABLE 2. Cumulative Data on Tumor Incidences in $CF_1$ Mice Untreated or Exposed to Phenobarbitone (PB)[a]

| Effective numbers of mice[b] | | Lymphomas | % Animals with | | |
|---|---|---|---|---|---|
| | | | Lung adenomas | Hepatomas | Osteomas |
| PB | 98 | 30.6 | 40.8 | 78.5 | 12.2 |
| Control | 44 | 43.2 | 45.5 | 27.3 | 11.4 |

[a] Adapted from Ponomarkov and Tomatis (20).
[b] Survivors at the time the first tumor was observed.

agent on nitrosamine-initiated cells in the mouse, similar to its action in rats. However, it has yet to be shown that this work is reproducible and, if so, it is unclear whether phenobarbital acts as a general hepatic neoplasm-enhancing agent in most or all strains of mice.

Based on previous work of phenobarbital effects in both rats and mice, it appears that phenobarbital, following nitrosamine initiation, holds promise for developing a hepatic initiation-promotion system in mice. However, two other systems have been presented in the literature that purportedly demonstrate two-stage hepatocarcinogenesis in mice. One system utilizes $LAF_1$ mice initiated with X-irradiation, followed by $CCl_4$ as the promoting agent. $CCl_4$ enhances hepatic tumor formation when given after, but not when given before, the initiating irradiation. However, it is not clear whether the promoting agent, carbon tetrachloride, is acting as a pure promoter or as a nonspecific stimulus to mitosis.

A possible system for studying two-stage transplacental liver carcinogenesis has been demonstrated in mice (23). Pregnant mice were treated with 2-AAF to initiate liver cells in the fetus. The newborn offspring were subsequently treated with intraperitoneal phorbol, which is a known tumor promoter in the skin. The offspring receiving phorbol had a slightly higher incidence of hepatic neoplasms compared to those receiving no phorbol. This system will require further refinement to be useful in studying initiation-promotion in hepatic carcinogenesis of the mouse.

In conclusion, there is a lack of adequately defined systems for studying initiation and promotion in mouse liver. However, based on the successful development of such systems in the rat, additional future efforts are warranted to develop initiation-promotion systems in mice. Subsequent use of these systems will be helpful in determining the biology of neoplastic development in the mouse liver. The information to be obtained from such studies is necessary

TABLE 3. Phenobarbital Enhancement of Dimethylnitrosamine-induced Liver Neoplasms[a]

| Treatment | Sex | Effective number of mice[b] | Number of mice with liver tumors | Average number of tumors/mouse |
|---|---|---|---|---|
| DMN + phenobarbital | M | 19 | 16 (84)[c] | 5.1 |
|  | F | 16 | 11 (69) | 3.3 |
| Total |  | 35 | 27 (77) | 4.3 |
| DMN | M | 13 | 4 (31) | 0.5 |
|  | F | 11 | 4 (36) | 0.3 |
| Total |  | 24 | 8 (33) | 0.4 |
| 0.9% NaCl + phenobarbital | M | 19 | 0 | 0 |
|  | F | 17 | 0 | 0 |
| Total |  | 36 | 0 | 0 |
| 0.9% NaCl | M | 14 | 0 | 0 |
|  | F | 7 | 0 | 0 |
| Total |  | 21 | 0 | 0 |

[a] Adapted from Uchida and Hirono (22).
[b] Mice surviving beyond 109 days after birth when the first liver neoplasm was found.
[c] Percent of surviving mice with liver neoplasms.

to explain differences in the incidence of hepatic neoplasms between rats and mice, as frequently noted in chronic bioassays.

REFERENCES

1. Rous P, Kidd JG: Conditional neoplasms and subthreshold neoplastic states: A study of the tar tumors of rabbits. J Exp Med 73:365-384, 1941.

2. Friedwald WF, Rous P: The initiating and promoting elements in tumor production. J Exp Med 80:101-125, 1944.

3. Berenblum J, Shubik P: The role of croton oil applications associated with a single painting of a carcinogen in tumor induction of the mouse's skin. Br J Cancer 1:379-391, 1947.

4. Berenblum J, Shubik P: An experimental study of the initiating stage of carcinogenesis and a reexamination of the somatic cell mutation theory of cancer. Br J Cancer 3:109-118, 1949.

5. Boutwell RK: The function and mechanism of promoters of carcinogenesis. Crit Rev Toxicol 2:419-443, 1974.

6. Farber E: Carcinogenesis: Cellular evolution as a unifying thread. Presidential address. Cancer Res 33:2537-2550, 1973.

7. Peraino C, Fy RJM, Staffeldt E: Reduction and enhancement by phenobarbital of hepatocarcinogenesis induced in the rat by 2-acetylaminofluorene. Cancer Res 31:1506-1512, 1971.

8. Peraino C, Staffeldt E, Haugen DA, Lombard LS, Stevens FJ, Fry RJM: Effects of varying the dietary concentration of phenobarbital on its enhancement of 2-acetylaminofluorene-induced hepatic tumorigenesis. Cancer Res 40:3268-3273, 1980.

9. Pitot HC, Sirica AE: The stages of initiation and promotion in hepatocarcinogenesis. Biochim Biophys Acta. 605:191-215, 1980.

10. Pitot HC, Barsness L, Goldsworthy T, Kitagawa T: Biochemical characterization of stages of hepatocarcinogenesis after a single dose of diethylnitrosamine. Nature(Lond) 271:456-458, 1978.

11. Ito N, Tatematsu M, Nakanishi K, Hasegawa R, Takano T, Imaida K, Ogiso T: The effects of various chemicals on the development of hyperplastic liver nodules in hepatectomized rats treated with N-nitrosodiethylamine or N-2-fluorenylacetamide. Gann 71:832-842, 1980.

12. Solt DB, Farber E: New principle for the analysis of chemical carcinogenesis. Nature 263:701-703, 1976.

13. Solt DB, Medline A, Farber E: Rapid emergence of carcinogen-induced hyperplastic lesions in a new model for the sequential analysis of liver carcinogenesis. Am J Pathol 88:595-618, 1977.

14. Cayama E, Tsuda H, Sarma DSR, Farber E: Initiation of chemical carcinogenesis requires cell proliferation. Nature(Lond) 275:60-62, 1978.

15. Shinozuka H, Sells MA, Katyal SL, Sell S, Lombardi B: Effects of choline-devoid diet on the emergence of γ-glutamyltranspeptidase-positive foci in the liver of carcinogen-treated rats. Cancer Res 39:2515-2521, 1979.

16. Abanobi SE, Lombardi B, Shinozuka H: Stimulation of DNA synthesis and cell proliferation in the liver of rats fed choline-devoid diet and their suppression by phenobarbital. Cancer Res 42:412-415, 1982.

17. Farber E: The sequential analysis of liver cancer induction. Biochim Biophys Acta 605:149-166, 1980.

18. Peraino C, Fry RJM, Staffeldt E: Enhancement of spontaneous hepatic tumorigenesis in C3H mice by dietary phenobarbital. J Natl Cancer Inst 51:1349-1350, 1973.

19. Kitagawa T, Watanabe R, Sugano H: Induction of γ-glutamyl transpeptidase activity by dietary phenobarbital in "spontaneous" hepatic tumors in C3H mice. Gann 71:536-542, 1980.

20. Ponomarkov V, Tomatis L: The effect of long-term administration of phenobarbitone in CF-1 mice. Cancer Lett 1:165-172, 1976.

21. Thorpe E, Walker AIT: The toxicology of dieldrin (HEOD). II. Comparative long-term oral toxicity studies in mice with dieldrin, DDT, phenobarbitone, β-BHC and γ-BHC. Food Cosmet Toxicol 11:433-442, 1973.

22. Uchida E, Hirono I: Effect of phenobarbital on induction of liver and lung tumors by dimethylnitrosamine in newborn mice. Gann 70:639-644, 1979.

23. Armuth V, Berenblum I: Possible two-stage transplacental liver carcinogenesis in C57BL/6 mice. Int J Cancer 20:292-295, 1977.

Chapter 9

# Mouse Hepatic Neoplasia: Differences among Strains and Carcinogens

**Boris H. Ruebner, M. Eric Gershwin, S. W. French, Earl Meierhenry, Pat Dunn, and Lucy S. Hsieh**

INTRODUCTION

The objective of this investigation was to study the biology of hepatic neoplasia in the mouse. Particular attention was paid to a comparison of the incidence and morphology of "spontaneous" hepatic neoplasms, with hepatic neoplasms induced by diethylnitrosamine and dieldrin. Diethylnitrosamine is active in multiple species and tissues, while dieldrin is a chlorinated hydrocarbon effective apparently only in the mouse liver (1,2). This comparison was carried out in males of three strains of mice: C57BL/6 mice are known to have a relatively low incidence of spontaneous hepatocellular tumors, C3H/He mice have a relatively high incidence (3), and the C57BL/6 x C3H/HeF1 (B6C3F1) hybrids are a popular strain in carcinogenesis testing.

MATERIALS AND METHODS

Weanling male C57BL/6, C3H/He, and C57BL/6 x C3H/He B6C3F1 hybrid mice from Charles River Laboratory, Wilmington, MA, were employed. The mice were housed in plastic cages with Absorbi-dri shavings, three per cage, and were given Purina Lab Chow and water ad libitum. The feed had been analyzed and found free of a panel of insecticides by Gulf South Research Institute, New Iberia, LA. Dieldrin, 98.5% pure, was obtained from Chemical Services, Inc., of West Chester, PA. The dieldrin was recrystallized from methanol by Dr. R. Krieger of the Department of Environmental Toxicology, University of California, Davis, CA, and was >99% pure, based

---

Supported in part by Public Health Service contract CP 65845 under the National Cancer Institute Carcinogenesis Testing program, Public Health Service grant ES 070554 for postdoctoral training in environmental pathology and by Veterans Administration research support.

upon analysis by electron-capture gas chromatography that showed
no impurities. The dieldrin was fed continuously at 10 ppm for 85
wk to 70 C57BL/6, 50 C3H/He, and 64 B6C3F1 mice. At 31 weeks after
the discontinuation of dieldrin (DIEL), blood samples were taken
for determination of serum dieldrin levels by electron-capture gas
chromatography. Diethylnitrosamine (DEN) was obtained from I.I.T.
Research Institute, Chicago, IL, and 12 µg were given
intraperitoneally (4,5) on days 0, 3, 9, and 15 to 61 C57BL/6, 59
C3H/He, and 65 B6C3F1 mice. A further 63 C57BL/6, 59 C3H/H3 and
64 B6C3F1 mice served as controls.

At 8 weeks after the initiation of the experiment and at intervals up
to 132 weeks of age, at least 3 mice in each group were sacrificed
and complete gross autopsies were performed. Tissues, including
the entire lungs and 3 slices from liver nodules and nonnodular
liver, were taken for light microscopy. In livers with multiple
nodules, up to 3 nodules were sampled and in livers without nodules,
3 slices were taken from the median or left hepatic lobes. For
light microscopy, the livers were studied in fresh-frozen cryostat
sections and in paraffin sections of material fixed in buffered
neutral formalin and stained with hematoxylin and eosin and by the
periodic acid Schiff reaction for glycogen. Fresh-frozen cryostat
sections were also employed for the demonstration of alkaline
phosphatase (6) (Alk Phos), glucose-6-phosphatase (G-6-Pase) (7),
succinic dehydrogenase (SDH) (8), and adenosine triphosphatase
(ATPase) (9). Liver nodules that contained trabecular areas
histologically were classified as hepatocellular carcinomas (10).
For electron microscopy, small blocks of liver tissue were fixed
by McDowell's method (11), embedded in Spurr's (12) low-viscosity
resin, and cut with a Porter Blum MT2 microtome. Ultrathin sections
were stained with uranyl acetate and lead citrate and photographed
in a Zeiss 9-S electron microscope.

Statistical analysis of data used the one-tailed test for the
difference of proportions.

RESULTS

Growth Curves and Liver Weights

The animals reached weights of approximately 40 grams after 1 year.
In all three strains the dieldrin animals were the heaviest
(Tables 1-3).

The livers of control animals weighed approximately 2.0 grams,
approximately 5.0% of the body weight. Livers with nodules weighed
up to 34% of the body weight (Table 1-3).

Gross Findings

Pathologic changes noted were virtually limited to the livers.
Small whitish spots and nodules were first seen after about 4 months
(Tables 1-3).

TABLE 1. Neoplasms in C57BL/6 Male Mice

| Age (weeks) | Mean body weight (grams) Control | DEN | Diel | Mean % liver weight Control | DEN | Diel | Number of mice with liver neoplasms Control | DEN | Diel |
|---|---|---|---|---|---|---|---|---|---|
| 9–14 | 39.5 | 28.1 | | | | | 0/6 | 0/6 | 0/6 |
| 15–25 | 31.2 | 36.4 | | | | | 1/6 | 1/6 | 2/6 |
| 26–42 | 32.7 | 31.9 | | | | | 0/6 | 0/6 | 0/6 |
| 43–46 | 37.6 | 37.9 | 34.3 | 5.7 | 6.3 | 7.7 | 0/3 | 0/3 | 1/3 |
| 47–52 | 38.0 | 36.8 | 42.3 | 6.7 | 5.3 | 7.3 | 0/3 | 0/3 | 2/3 |
| 53–60 | 45.3 | 32.1 | 41.7 | 6.0 | 6.3 | 6.3 | 1/3 | 1/3 | 2/3 |
| 61–67 | 34.1 | 36.1 | 47.1 | 4.7 | 5.0 | 7.7 | 1/3 | 0/3 | 3/3 |
| 68–70 | 36.3 | 33.5 | 53.3 | 5.0 | 5.0 | 10.0 | 1/3 | 1/3 | 2/3 |
| 71–75 | 39.5 | 35.4 | 37.2 | 4.7 | 5.0 | 7.3 | 1/3 | 1/3 | 3/3 |
| 76–78 | 32.0 | 33.8 | 45.0 | 5.7 | 6.0 | 7.7 | 1/3 | 2/3 | 2/3 |
| 79–86 | 30.5 | 32.4 | 33.2 | 5.7 | 8.3 | 11.3 | 0/3 | 0/3 | 3/3 |
| 87–88 | 28.9 | 22.2 | 43.5 | 5.7 | 5.0 | 13.3 | 1/3 | 2/3 | 3/3 |
| 89–97 | 29.2 | 24.4 | 32.1 | 5.0 | 5.3 | 9.0 | 1/3 | 0/3 | 2/3 |
| 98–102 | 28.1 | 27.8 | 35.5 | 5.0 | 7.0 | 10.0 | 0/3 | 1/3 | 3/3 |
| 103–114 | 29.1 | 32.0 | 33.4 | 5.9 | 6.7 | 9.3 | 1/3 | 3/3 | 1/3 |
| 115–123 | | | 30.1 | 7.7 | 5.8 | 10.8 | 1/3 | 1/3 | 3/3 |
| 124–127 | | | 30.8 | 6.2 | 7.5 | 8.8 | 1/3 | 2/3 | 3/4 |
| 128–132 | | | 26.8 | 6.3 | 9.3 | 10.8 | 0/3 | 0/1 | 6/9 |
| Average | 34.2 | 32.0 | 37.8 | | | | 11/63 (17%) | 14/61 (23%) | 60/70 (57%) |

TABLE 2. Neoplasms in C3H/He Male Mice

|  | Mean body weight (grams) |  |  | Mean % liver weight |  |  | Number of mice with liver neoplasms |  |  |
|---|---|---|---|---|---|---|---|---|---|
| Age (weeks) | Control | DEN | Diel | Control | DEN | Diel | Control | DEN | Diel |
| 8-16 |  |  |  |  |  |  | 0/6 | 0/6 | 0/6 |
| 17-25 |  |  |  |  |  |  | 0/6 | 0/6 | 0/6 |
| 26-40 |  |  |  |  |  |  | 0/6 | 3/6 | 2/6 |
| 41-48 | 40.3 | 38.8 | 42.3 | 5.7 | 6.0 | 7.0 | 2/6 | 4/6 | 4/6 |
| 49-58 | 41.5 | 32.7 | 43.9 | 5.3 | 7.3 | 8.2 | 1/3 | 1/3 | 3/3 |
| 59-64 | 39.7 | 40.3 | 41.9 | 4.7 | 5.7 | 7.7 | 0/3 | 3/3 | 3/3 |
| 65-68 | 38.6 | 40.4 | 40.3 | 5.7 | 10.7 | 10.3 | 2/3 | 3/3 | 3/3 |
| 69-72 | 34.7 | 33.2 | 38.7 | 4.3 | 9.0 | 17.3 | 1/3 | 2/3 | 3/3 |
| 73-100 | 31.5 | 29.6 | 34.8 | 5.3 | 12.7 | 18.3 | 2/3 | 3/3 | 3/3 |
| 101-111 | 32.5 | 30.8 | 32.4 | 8.0 | 8.5 | 13.9 | 13/20 | 18/20 | 11/11 |
| Average | 37.0 | 35.1 | 39.2 |  |  |  | 21/59 (35%) | 35/59 (59%) | 32/50 (64%) |

TABLE 3. Neoplasms in B6C3F1 Male Mice

| Age (weeks) | Mean body weight (grams) Control | Mean body weight (grams) DEN | Mean body weight (grams) Diel | Mean % liver weight Control | Mean % liver weight DEN | Mean % liver weight Diel | Number of mice with liver neoplasms Control | Number of mice with liver neoplasms DEN | Number of mice with liver neoplasms Diel |
|---|---|---|---|---|---|---|---|---|---|
| 15-20 | 36.0 | 34.2 | 38.5 | 5.7 | 5.3 | 5.7 | 0/6 | 1/6 | 0/6 |
| 21-24 | 30.5 | 31.1 | 35.1 | 4.0 | 4.3 | 5.0 | 0/3 | 1/3 | 0/3 |
| 25-31 | 37.9 | 45.0 | 41.0 | 5.3 | 5.0 | 6.7 | 0/3 | 1/3 | 2/3 |
| 32-38 | 38.6 | 35.0 | 36.3 | 4.7 | 4.7 | 6.0 | 0/3 | 0/3 | 1/3 |
| 39-45 | 42.5 | 45.2 | 45.3 | 4.7 | 4.3 | 7.0 | 0/3 | 0/3 | 1/3 |
| 46-49 | 37.2 | 48.1 | 43.6 | 5.7 | 5.0 | 6.7 | 0/3 | 1/3 | 3/3 |
| 50-53 | 34.1 | 41.5 | 43.4 | 5.0 | 5.0 | 8.3 | 1/3 | 2/3 | 2/3 |
| 54-56 | 38.3 | 46.5 | 46.6 | 6.7 | 4.7 | 7.3 | 0/3 | 2/3 | 3/3 |
| 57-62 | 37.7 | 48.9 | 44.9 | 5.0 | 6.0 | 7.7 | 1/3 | 1/3 | 2/3 |
| 63-64 | 35.2 | 32.0 | 43.9 | 5.0 | 4.3 | 13.3 | 0/3 | 1/3 | 3/3 |
| 65-75 | 37.7 | 38.8 | 41.0 | 7.3 | 8.5 | 20.0 | 1/3 | 1/3 | 3/3 |
| 76-92 | 35.8 | 36.3 | 37.2 | 6.3 | 6.4 | 15.4 | 1/3 | 1/3 | 3/3 |
| 93-98 | 30.8 | 34.1 | 38.0 | 6.3 | 10.7 | 27.5 | 3/12 | 4/9 | 8/9 |
| 99-106 | 33.3 | 31.5 | 34.2 | 6.8 | 5.8 | 25.8 | 0/3 | 2/3 | 3/3 |
| 107-110 | 32.3 | 31.6 | 37.6 | 6.4 | 14.6 | 27.3 | 0/3 | 2/3 | 3/3 |
| 111-116 | 34.0 | 29.4 | 30.9 | 6.6 | 9.5 | 34.5 | 0/2 | 3/3 | 3/3 |
| 117-119 |  | 33.6 | 37.3 |  | 10.0 | 25.1 | 1/2 | 3/3 | 3/3 |
| 120-124 |  |  |  |  |  |  | 1/2 | 0/0 | 3/3 |
| 124-132 |  |  |  |  |  |  |  | 0/0 | 3/3 |
| Average | 35.7 | 37.8 | 39.7 |  |  |  | 8/64 (12.5%) | 32/65 (50%) | 48/64 (75%) |

Dieldrin-fed animals of all three strains had a higher incidence of hepatic tumors than those that had received DEN. DEN animals in turn had a higher incidence than controls (Table 4). The mean number of neoplasms per liver was also higher in the dieldrin animals than in the DEN animals, which had more tumors per liver than the controls (Figs. 1-3). The mean diameters of the hepatic neoplasms however, were not very different in the three groups (Figs. 4-6).

As expected, the incidence of "spontaneous" hepatic tumors among C3H mice was higher than that of the C57 (Table 4). Dieldrin and diethylnitrosamine-treated C3H mice also had more tumors than similarly treated C57 mice. The B6C3F1 hybrids, surprisingly, resembled the C57 parent strain by having a low incidence of tumors in the untreated control animals and the C3H parent strain by having a high incidence of tumors when given dieldrin or diethylnitrosamine (Table 4). The mean numbers of tumors per liver (Figs. 1-3) and their mean diameters (Figs. 4-6) were higher in the B6C3F1 hybrids than in either parent strain.

Light Microscopy

The livers of the DEN animals showed no diffuse toxic changes. In the dieldrin animals, however, there was marked swelling of hepatocytes in central zones associated with some nuclear atypia. Bile duct or oval cell proliferation was not detected in any of the animals.

Small microscopic foci of putative preneoplastic hepatocytes were seen in all groups. However, they were so scanty that no conclusions as to either their relative frequency in different groups or their histochemistry could be drawn. These foci did not appear to have a consistent localization in the lobule. Most foci were composed principally of clear hepatocytes.

TABLE 4. Number of Animals with Hepatic Tumors in Different Mouse Strains Treated with Dieldrin or DEN

| Treatment | C57BL/6 Strain | C3H/He Strain | B6C3F1 Strain |
|---|---|---|---|
| Dieldrin | 40/70 (57.0%) | 32/50 (64.0%) | 48/64 (75.0%) |
| DEN | 14/61 (23.0%) | 35/59 (59.0%) | 32/65 (50.0%) |
| Controls | 11/63 (17.0%) | 21/59 (35.0%) | 8/64 (13.0%) |

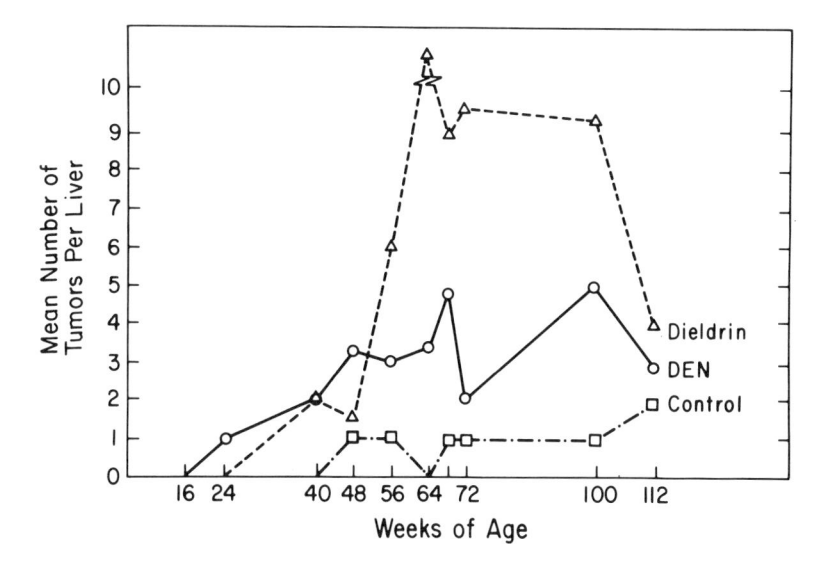

DIFFERENCES AMONG STRAINS AND CARCINOGENS 121

FIGURE 1. Multiplicity of liver tumors in C57 mice with time.

The nodules consisted of plates or sheets of hepatocytes without
any lobular arrangement. Usually, the hepatocytes in the nodules
were somewhat larger than those outside the nodules. The hepatocytes
in the nodules differed in their morphology. Some had clear

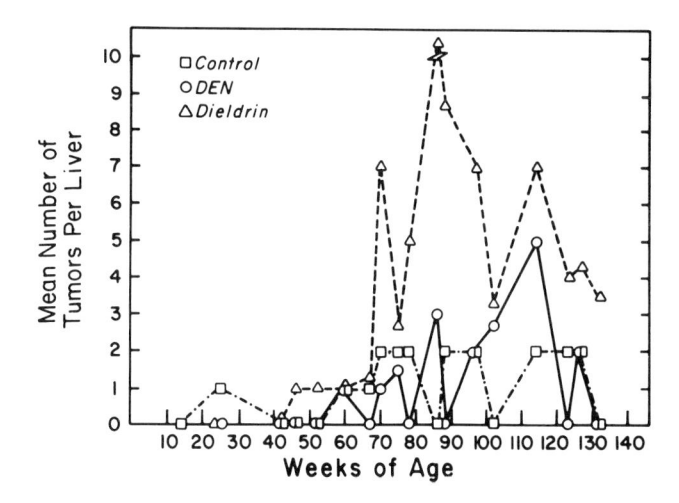

FIGURE 2. Multiplicity of liver tumors in C3H mice with time.

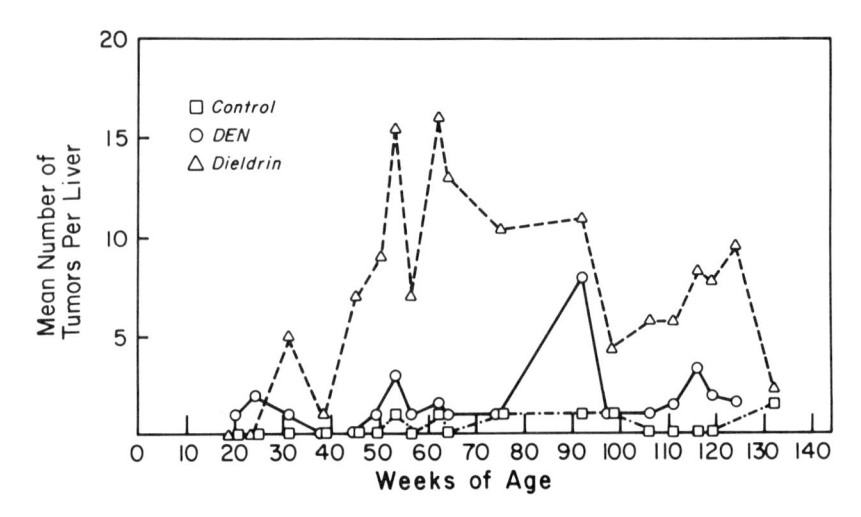

FIGURE 3. Multiplicity of liver tumors in B6C3F1 mice with time.

cytoplasm (Fig. 7) and resembled those seen in the foci described above. Others had basophilic (Fig. 7) or eosinophilic (Fig. 8) cytoplasm. Often the nodules consisted of more than one of these cell types. Nodules that contained clear cells frequently also contained basophilic or eosinophilic cells. In mixed nodules, clear cells tended to be located centrally and basophilic cells on the periphery (Fig. 7). Table 5 shows the frequency of predominantly basophilic and predominantly eosinophilic tumors in the different strains and treatment groups. Multiple tumors were counted and histologically studied independently as far as possible. Table 5 shows that, irrespective of treatment, tumors in the C3H strain were much more frequently basophilic than eosinophilic. However,

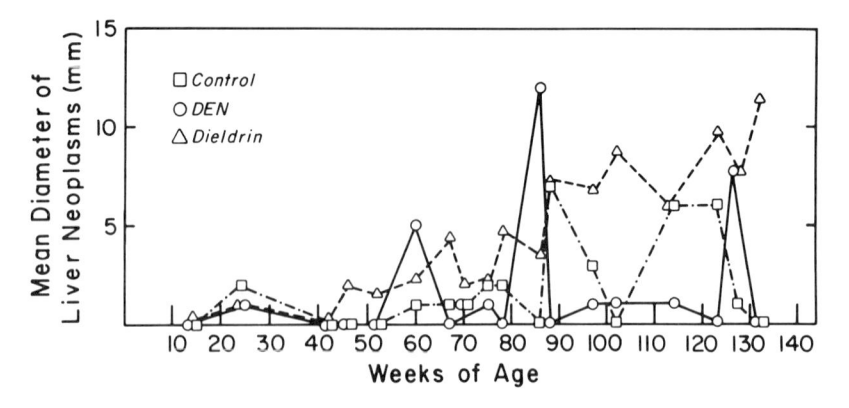

Figure 4. Growth of liver tumors in C57 mice with time.

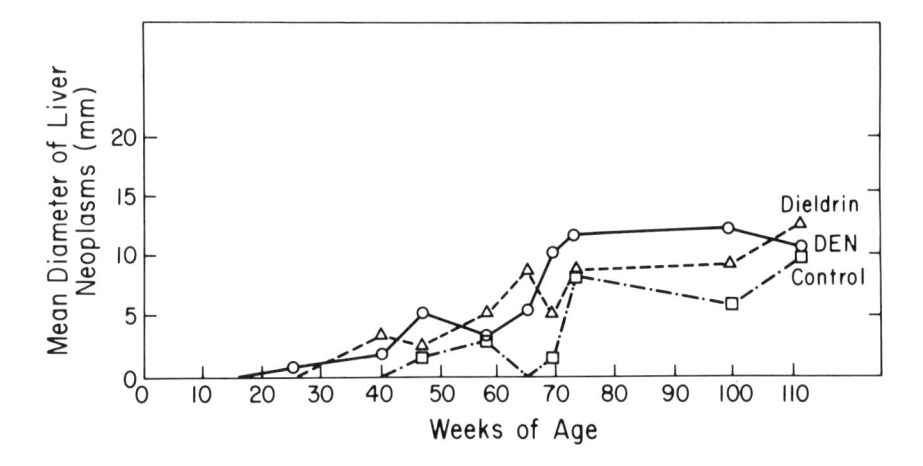

FIGURE 5. Growth of liver tumors in C3H mice with time.

the C57 and B6C3F1 hybrids had significantly more eosinophilic than basophilic tumors. This was particularly striking in the dieldrin-treated animals.

Hepatocellular carcinoma (HCC) was diagnosed when entire hepatocellular nodules or significant parts of nodules showed abnormal trabecular hepatic plate architecture and cytologic atypia (Fig. 9). The incidence of HCC among the three strains and among the different treatment groups (Table 6) resembled that of all hepatic neoplasms (Table 4). Animals given dieldrin had a higher incidence of HCC than those that had DEN, and these in turn had more HCC than controls. As expected, the C3H strain had a higher incidence of HCC than the C57. The B6C3F1 hybrids resembled their C3H parent strain in the high incidence of HCC induced by dieldrin and DEN,

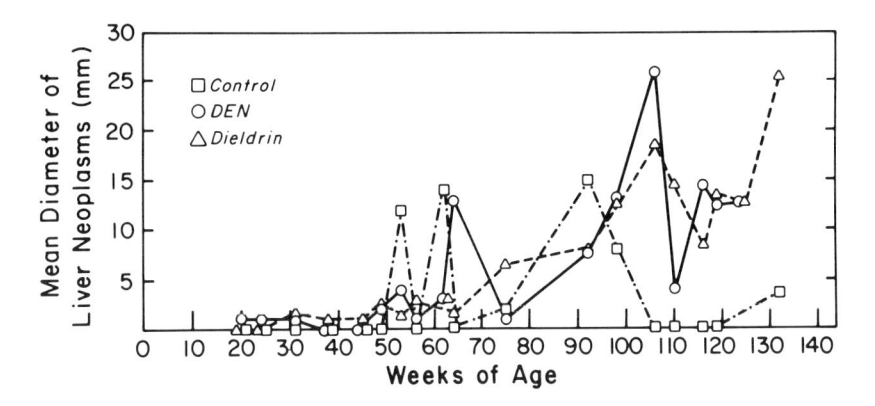

FIGURE 6. Growth of liver tumors in B6C3F1 mice with time.

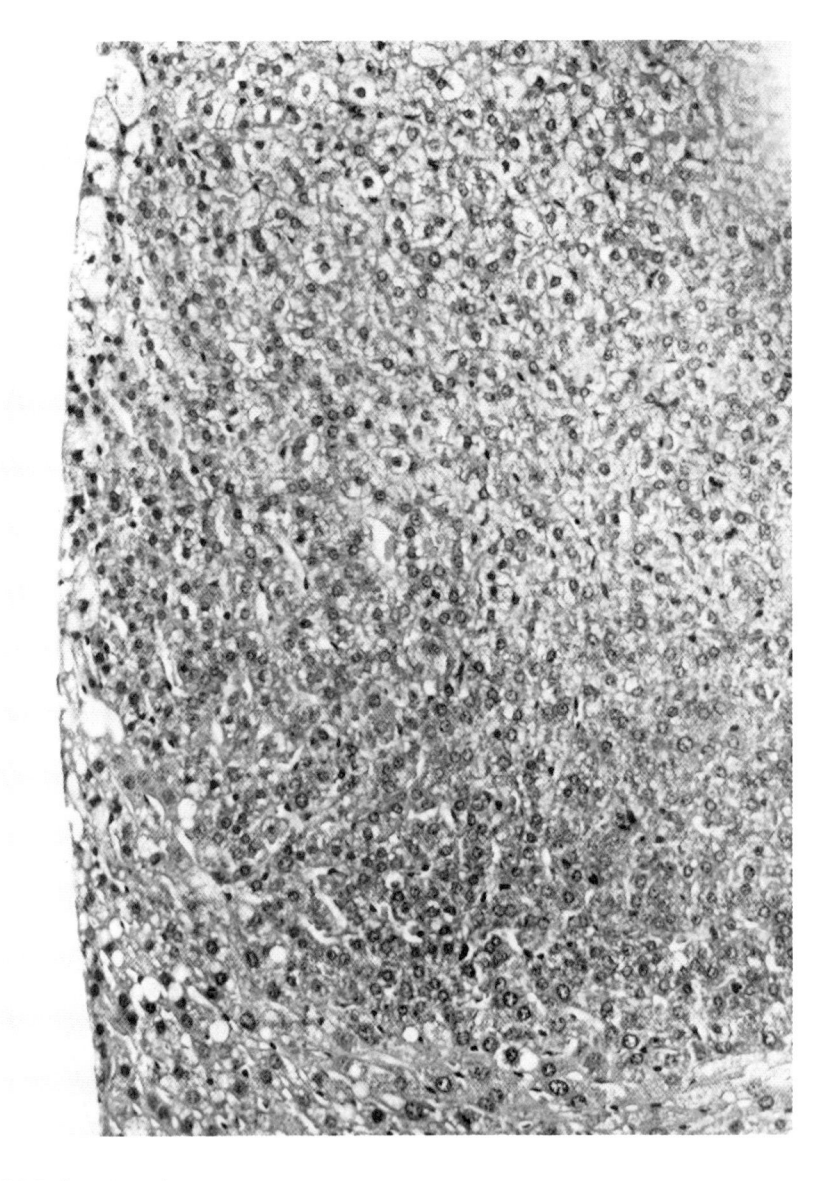

FIGURE 7. An adenoma composed of clear hepatocytes in the center (upper right) and basophilic cells at the periphery. Some compressed normal liver is seen at the bottom left. DEN-treated C3H mouse. Hematoxylin and eosin, X200. Reprinted by permission, from Ruebner et al. (17).

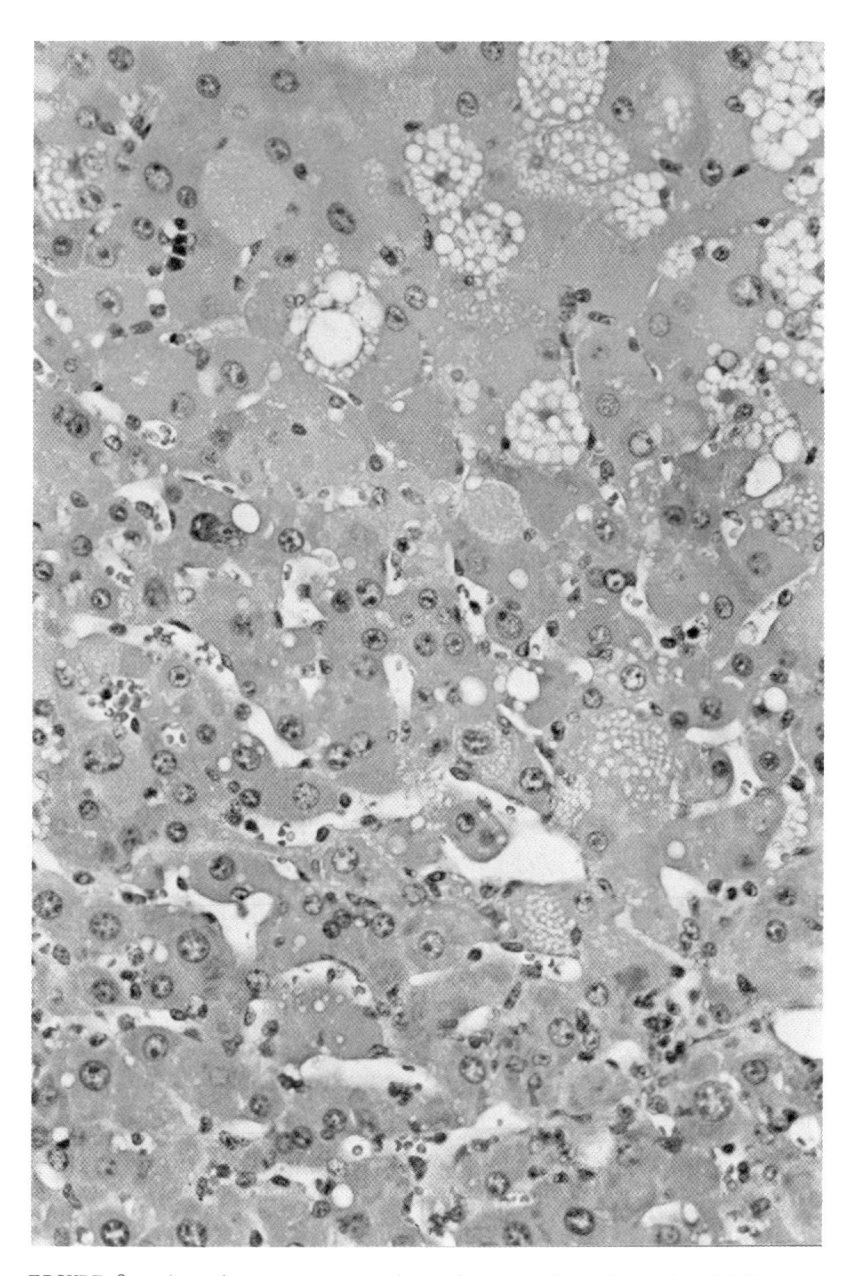

FIGURE 8. An adenoma composed predominantly of eosinophilic hepatocytes. Scattered through the adenoma are clear hepatocytes containing many small lipid droplets. Some compressed normal hepatocytes are seen at the bottom left. Dieldrin-treated B6C3F1 mouse. Hematoxylin and eosin, X200.

TABLE 5. Basophilic and Eosinophilic Hepatic Neoplasms in Control Mice and in Mice Treated with DEN and Dieldrin

|        | DEN        |              | Dieldrin   |              | Control    |              |
|--------|------------|--------------|------------|--------------|------------|--------------|
| Strain | Basophilic | Eosinophilic | Basophilic | Eosinophilic | Basophilic | Eosinophilic |
| C3H    | 26         | 8            | 24         | 11           | 11         | 3            |
| C57    | 3          | 5            | 6          | 27           | 2          | 2            |
| B6C3F1 | 13         | 12           | 1          | 34           | 3          | 1            |

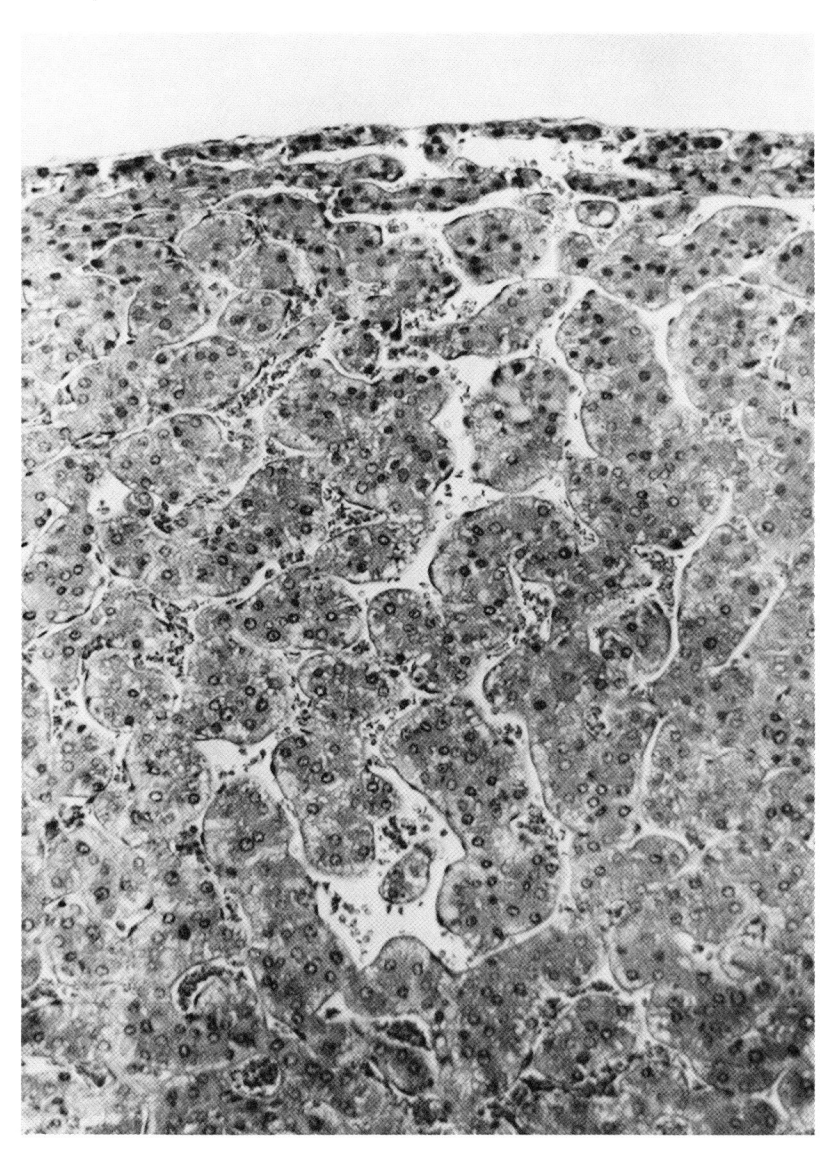

FIGURE 9. Trabecular hepatocellular carcinoma, DEN-treated C3H mouse. Hematoxylin and eosin, X200. Reprinted by permission, from Ruebner et al. (17).

TABLE 6. Number of Animals with Hepatocellular Carcinomas in Different Mouse Strains Treated with Dieldrin or DEN

| Treatment | C57BL/6J | C3H/He | B6C3F1 |
|---|---|---|---|
| Dieldrin | 21/70 (30.0%) | 19/50 (38.0%) | 26/64 (40.6%) |
| DEN | 3/61 (4.0%) | 16/59 (27.1%) | 13/65 (20.0%) |
| Controls | 0 (0.0%) | 6/59 (10.1%) | 3/64 (4.7%) |

and their C57 parent strain in the low incidence of HCC among controls.

Our finding of Mallory bodies (MBs) (Fig. 10) in the hepatocellular neoplasms of the dieldrin-treated animals was unexpected. These cytoplasmic inclusions were detected in all three strains but were significantly less frequent in the C3H animals than in the C57 strain or the B6C3F1 hybrids (Table 7). MBs were found in both benign and malignant hepatocellular tumors but were only rarely seen in mice without hepatocellular neoplasms.

Enzyme histochemistry was employed to visually compare the activity of each nodule with adjacent nonnodular liver (Fig. 11). Different hepatic tumors in a single animal were considered independent tumors. Table 8 shows that, with all enzymes, a significantly greater proportion of carcinomas than adenomas had increased enzyme activity in comparison with nonnodular liver. The glycogen content of the tumors, however, had the opposite pattern, since a significantly greater proportion of carcinomas showed decreased glycogen staining.

Analysis of histochemical reactions by treatment groups (Table 9) indicates that a significantly greater proportion of nodules in the dieldrin animals had increased activities for Alk Phos, G-6-Pase, and ATPase compared to nodules of control groups. This was true for both benign and malignant tumors.

Electron Microscopy

The clear cells in the nodules differed from the hepatocytes of controls in several respects. The clear areas of the cytoplasm of these cells were occupied either by glycogen granules (Fig..12), fat droplets (Fig. 13), or a combination of both of these.

The basophilic cells in the nodules differed from the clear cells principally in showing a striking increase in rough endoplasmic reticulum (Fig. 14). Hepatocytes, intermediate between clear and basophilic cells, were observed quite frequently. In some basophilic hepatocytes the rough endoplasmic reticulum was dilated and contained a uniform appearing secretory product. Occasional basophilic hepatocytes had a markedly increased number of microbodies.

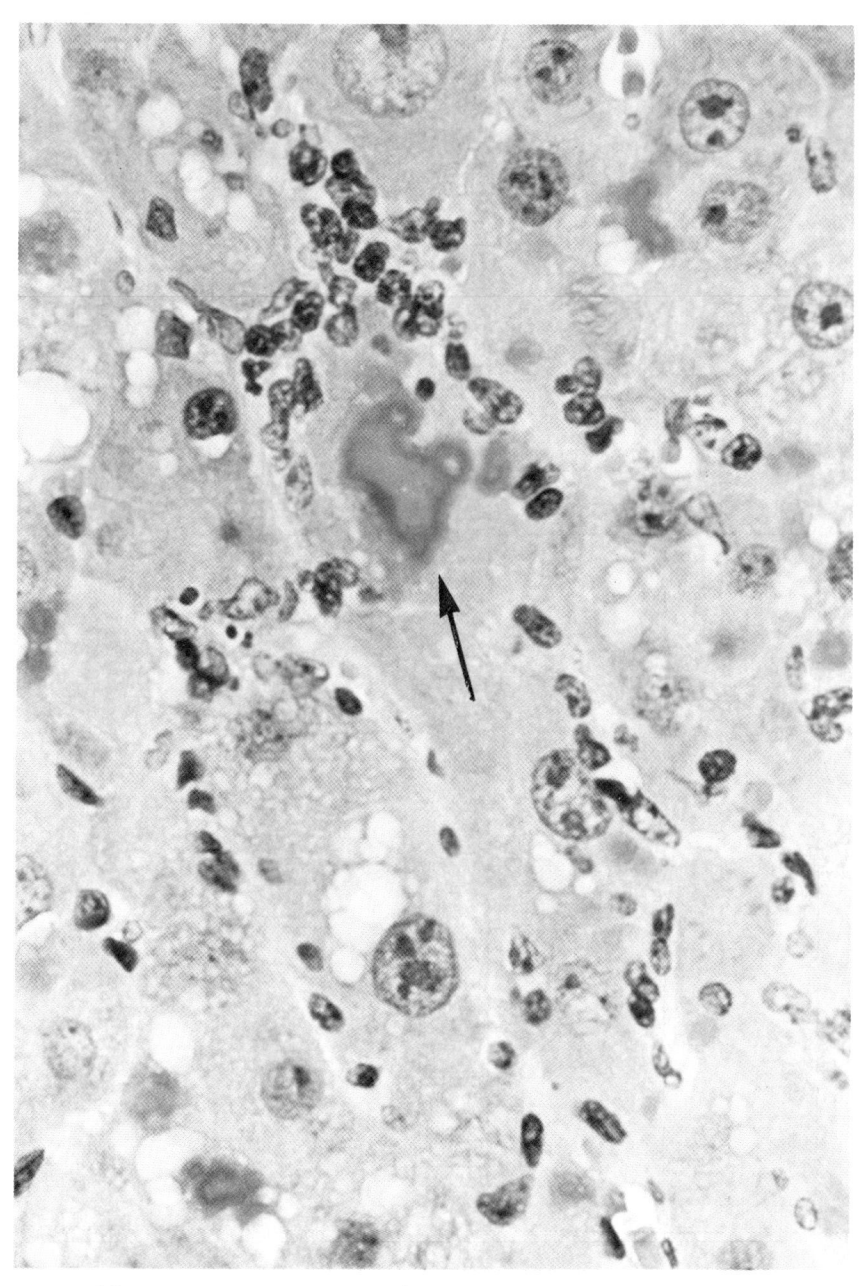

FIGURE 10. Large Mallory body (arrow) surrounded by inflammatory cells in an eosinophilic adenoma. Dieldrin-treated C57 mouse. Hematoxylin and eosin, X800. Reprinted by permission, from Meierhenry et al. (12).

TABLE 7. Number of Animals with Mallory Bodies (MBs) among Dieldrin-treated Mice

| Strain | Incidence of MBs in entire group | Incidence of MBs in mice with hepatic tumors[a] | Incidence of MBs in mice with hepatocellular carcinoma | Incidence of MBs in mice without tumors |
|---|---|---|---|---|
| C57BL/6 | 27/71 (38%) | 26/41 (63%)[b] | 20/21 (95%)[c] | 1/30 (3%) |
| C3H/He | 12/38 (32%) | 12/29 (41%) | 10/19 (53%) | 0/9 (0%) |
| B6C3F1 | 39/62 (63%)[d] | 36/44 (82%)[c,e] | 23/26 (88%)[c] | 3/18 (17%) |

[a] Includes both benign and malignant tumors.
[b] p < .05 compared with C3H/H3.
[c] p < .10 compared with C3H/He.
[d] p < .10 compared with C3H/He or C57BL/6.
[e] p < .05 compared with C57BL/6.

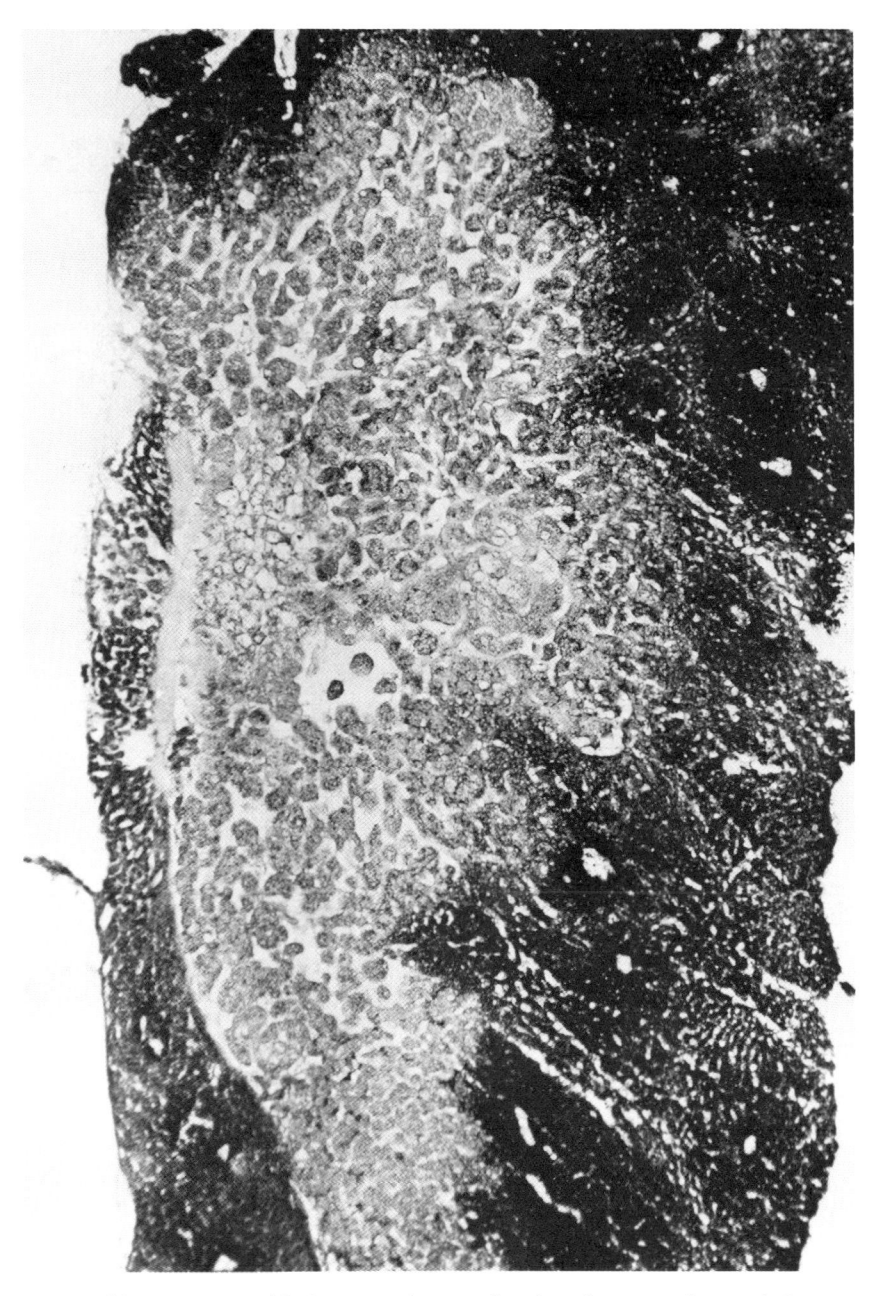

FIGURE 11. Hepatocellular carcinoma showing decreased succinic dehydrogenase reaction. DEN-treated B6C3F1 mouse. X100.

TABLE 8. Histochemical Reactions of Hepatic Adenomas and Carcinomas

|  | Alkaline phosphatase |
|---|---|
| Adenomas | 35↑   38↓   18 NC |
| Carcinomas | 79↑   24↓   13 NC |

|  | Glucose-6-phosphatase |
|---|---|
| Adenomas | 21↑   52↓   18 NC |
| Carcinomas | 60↑   47↓   24 NC |

|  | Succinic dehydrogenase |
|---|---|
| Adenomas | 24↑   45↓   22 NC |
| Carcinomas | 48↑   49↓   18 NC |

|  | Adenosine triphosphatase |
|---|---|
| Adenomas | 7↑   3↓   82 NC |
| Carcinomas | 40↑   10↓   64 NC |

|  | Glycogen content |
|---|---|
| Adenomas | 20↑   8↓   65 NC |
| Carcinomas | 21↑   37↓   59 NC |

Numbers represent numbers of tumors.
↑ = Increased histochemical reaction in tumors when compared to nonneoplastic liver.
↓ = Decreased histochemical reaction in tumors when compared to nonneoplastic liver.
NC = no change.

TABLE 9. Histochemical Reactions of Tumors Induced with Different Carcinogens

| Treatment | Adenomas | Carcinomas |
|---|---|---|
| | Alkaline phosphatase | |
| Dieldrin | 22↑ 11↓ 8 NC | 54↑ 9↓ 5 NC |
| DEN | 9↑ 20↓ 7 NC | 17↑ 9↓ 5 NC |
| Control | 5↑ 7↓ 3 NC | 8↑ 6↓ 3 NC |
| | Glucose-6-phosphatase | |
| Dieldrin | 12↑ 22↓ 7 NC | 41↑ 27↓ 15 NC |
| DEN | 6↑ 23↓ 6 NC | 12↑ 14↓ 5 NC |
| Control | 3↑ 7↓ 5 NC | 7↑ 6↓ 4 NC |
| | Succinic dehydrogenase | |
| Dieldrin | 11↑ 18↓ 13 NC | 27↑ 29↓ 11 NC |
| DEN | 10↑ 20↓ 5 NC | 15↑ 14↓ 2 NC |
| Control | 3↑ 7↓ 4 NC | 6↑ 6↓ 5 NC |
| | Adenosine triphosphatase | |
| Dieldrin | 5↑ 0↓ 37 NC | 26↑ 7↓ 34 NC |
| DEN | 1↑ 3↓ 31 NC | 13↑ 2↓ 15 NC |
| Control | 1↑ 0↓ 14 NC | 1↑ 1↓ 15 NC |
| | Glycogen content | |
| Dieldrin | 6↑ 3↓ 35 NC | 13↑ 22↓ 34 NC |
| DEN | 8↑ 4↓ 22 NC | 5↑ 11↓ 15 NC |
| Control | 6↑ 1↓ 8 NC | 3↑ 4↓ 10 NC |

Numbers represent numbers of tumors.
↑ = Increased histochemical reaction in tumors when compared to nonneoplastic liver.
↓ = Decreased histochemical reaction in tumors when compared to nonneoplastic liver.
NC = no change.

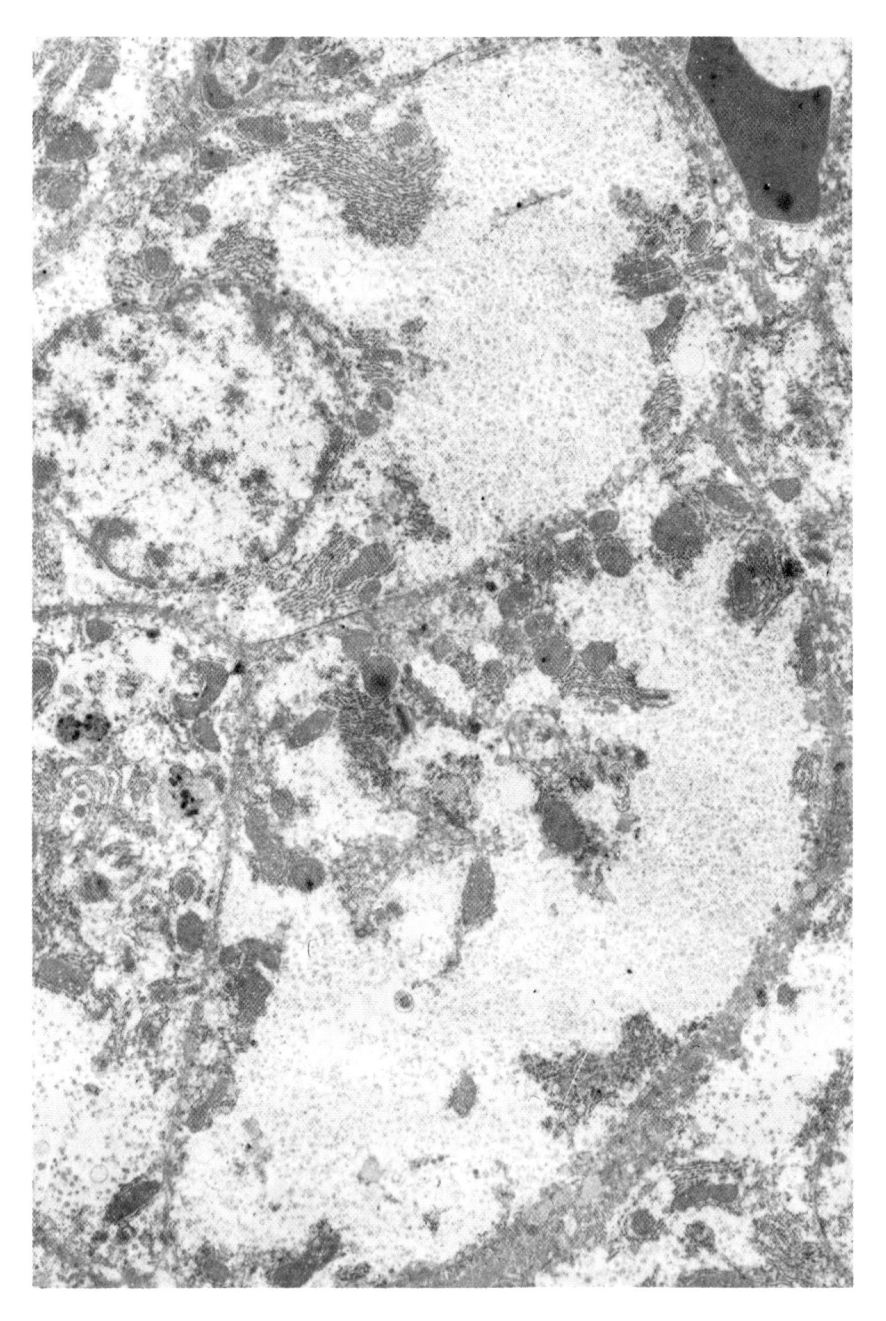

FIGURE 12. A clear hepatocyte in an adenoma. The glycogen area is greatly enlarged at the expense of other organelles. Dieldrin-treated C3H mouse. Electron micrograph, X7000. Reprinted by permission, from Ruebner et al. (17).

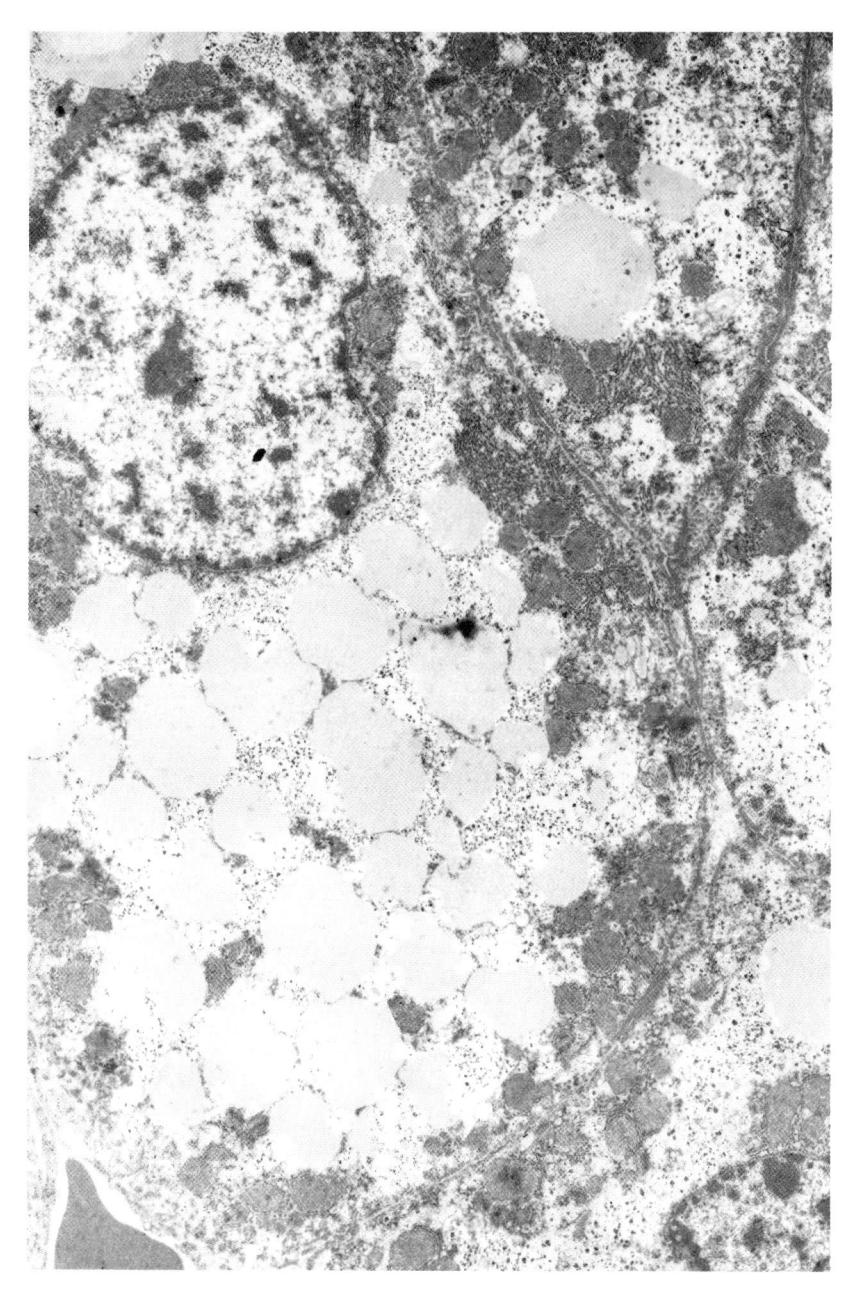

FIGURE 13. A different clear hepatocyte in the same nodule as Fig. 12. Most of the cytoplasm is composed of lipid droplets, with some scattered glycogen granules. Electron micrograph, X8000. Reprinted by permission, from Ruebner et al. (17).

FIGURE 14. A basophilic hepatocyte from an adenoma. Rough endoplasmic reticulum is the predominant cytoplasmic organelle. Some glycogen granules and lipid droplets are still present. DEN-treated C3H mouse. Electron micrograph, X10,000. Reprinted by permission, from Ruebner et al. (17).

Eosinophilic hepatocytes in the nodules differed from clear and basophilic hepatocytes by having mitochondria and smooth endoplasmic reticulum (SER) as the predominant cytoplasmic organelles. Some hepatocytes contained predominantly SER (Fig. 15), others contained predominantly mitochondria (Fig. 16), and yet others showed a proliferation of both of these organelles. We were unable to predict the ultrastructure of eosinophilic nodules from their light microscopy. Some eosinophilic hepatocytes, like some basophilic hepatocytes, showed a markedly increased number of microbodies.

The cytoplasm of cells in the hepatocarcinomatous nodules varied from clear to basophilic and eosinophilic. Ultrastructurally, the cells composing the trabecular carcinomatous areas were notable by having occasional basement membranes in the space of Disse. Focal separations of the lateral cell membranes with the development of microvilli were also conspicuous. Dilation of the rough endoplasmic reticulum and accumulation of secretions in the cisternae were also more marked in the hepatocarcinomas than in the benign nodules.

DISCUSSION

Dieldrin (Diel), a carcinogen effective apparently only in the mouse liver (1,2), produced a higher incidence of hepatic tumors than diethylnitrosamine (DEN). This was true in all three strains and for both benign and malignant tumors (Table 4 and 6). Evidently, the dose of DEN that we selected was a relatively weak one for the animals we employed. A persistent problem in experimental carcinogenesis with mice is the development of "spontaneous" hepatocelluar tumors in untreated control animals. Such neoplasms were observed in all strains, and their incidence was highest in C3H mice (Table 4). Both C57 mice and the hybrids produced by crossing C3H and C57 (B6C3F1) mice had a considerably lower incidence of tumors in controls. Surprisingly, the B6C3F1 hybrids were as susceptible to hepatic carcinogenesis as the C3H parent strain. This suggests that, in different strains, the incidence of hepatic tumors in control mice is not necessarily correlated with susceptibility to hepatic carcinogenesis. Ideally, a mouse strain employed in hepatic carcinogenesis studies should have no hepatic tumors in controls, but a high susceptibility to carcinogenesis. The C57 and B6C3F1 strains both approached zero incidence in controls, as far as hepatocellular tumors were concerned, and the C57 strain actually had a zero incidence of hepatocellular carcinomas in controls (Table 6). The B6C3F1 hybrids were, however, considerably more susceptible to hepatic carcinogenesis than the parent C57 strain. This indicates that the B6C3F1 hybrids, a popular strain in carcinogenesis testing, are actually quite suitable for such tests. A similar conclusion was recently reached by Dragani et al. (13). However, we are not aware of any evidence that would indicate that the B6C3F1 strain is necessarily the best of the mouse strains currently available for studies on hepatocarcinogenesis.

The results just discussed, like most current carcinogenesis testing, are based on a comparison of animals positive for tumor in different groups. These groups were composed of males of

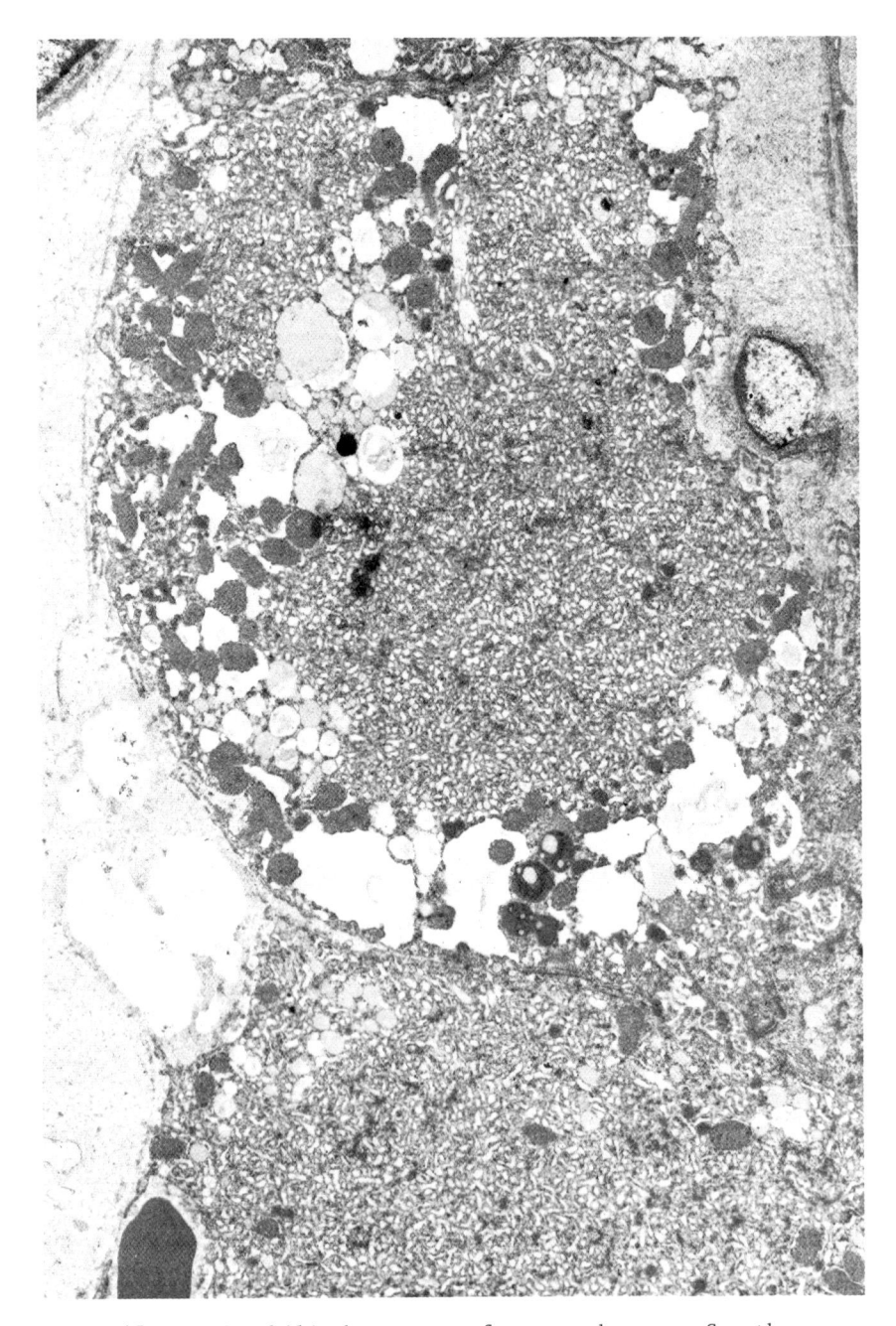

FIGURE 15. Eosinophilic hepatocyte from an adenoma. Smooth endoplasmic reticulum is the predominant cytoplasmic organelle. The other organelles are displaced towards the periphery of the cell. Dieldrin-treated C57 mouse. Electron micrograph, X6300.

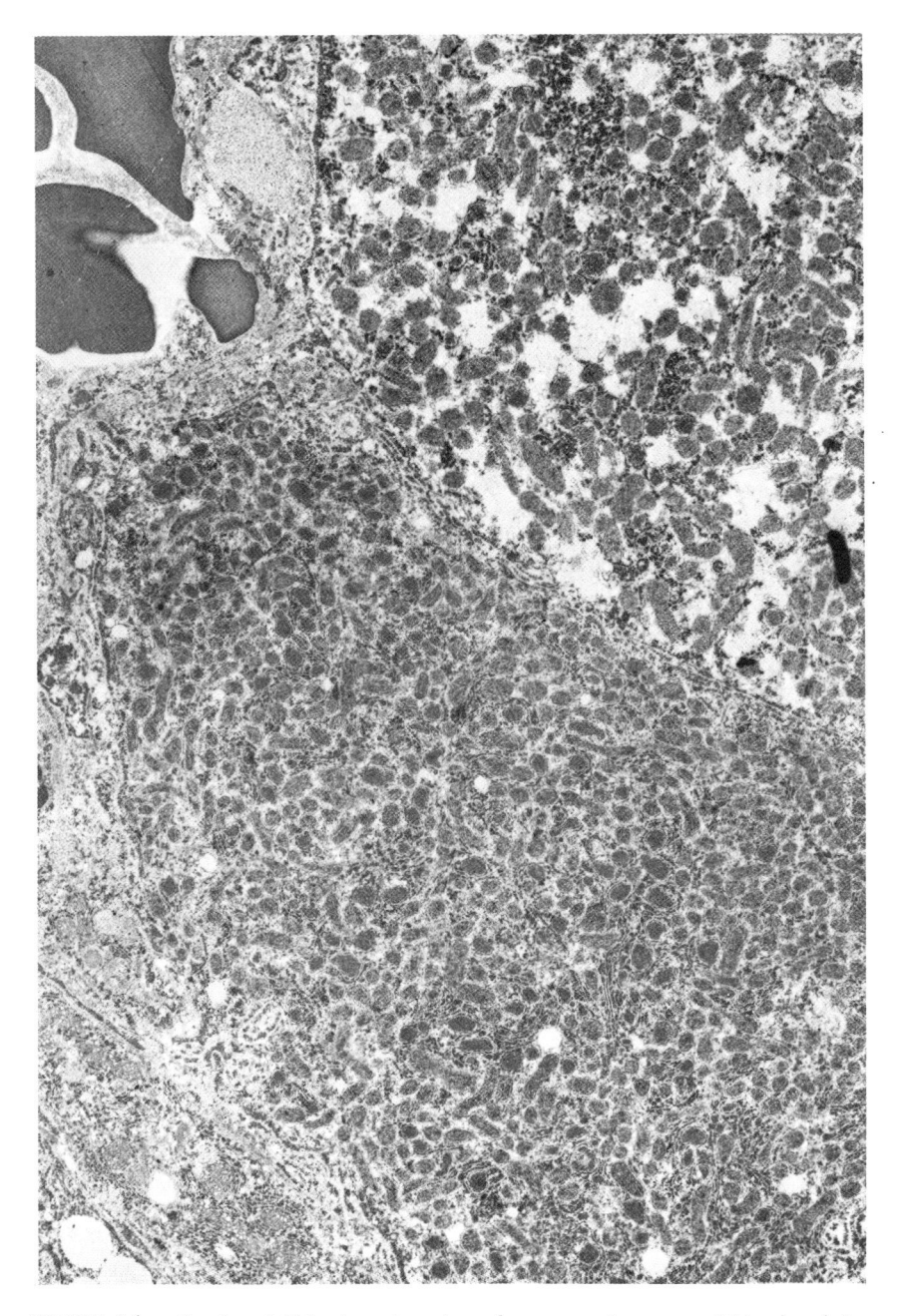

FIGURE 16. Eosinophilic hepatocytes from an adenoma. Mitochondria are the predominant cytoplasmic organelle. Dieldrin-treated C57 mouse. Electron micrograph, X6300.

different strains treated with different carcinogens. Counting the number of tumors per animal and measuring tumor size provides additional parameters that may be used to compare the carcinogenicity of different compounds. In our experiment the mean number of tumors per liver (Tables 1-3) proved to be a useful independent parameter for this purpose. The mean diameter of tumors per liver (Tables 4-6) proved to be less discriminating. The principal problem with both parameters concerns animals with tumors that are confluent and therefore cannot be counted or measured accurately.

Scanty microscopic foci of putatively preneoplastic cells were observed among all groups in this experiment. The reason for the scarcity of the foci in our animals is not clear since foci appear to be quite common in other mouse systems (4,13). The similarity of the clear cells in the foci to the clear cells in the nodules suggests that foci may develop into nodules.

Clear and basophilic cells were characteristic of the hepatic nodules among C3H animals (15). Ultrastructurally, the clear areas of these tumor cells contained glycogen granules and lipid droplets in varying proportions. Basophilic cells were characterized by proliferation of rough endoplasmic reticulum. Hepatocytes, intermediate between clear and basophilic hepatocytes, were quite frequent, suggesting that the former might be precursors of the latter. Nodules composed predominantly of eosinophilic cells were characteristic of the C57 and B6C3F1 strains, particularly among groups fed dieldrin (Table 5). Ultrastructurally, these eosinophilic hepatocytes differed considerably from the clear and basophilic hepatocytes. The predominant cytoplasmic organelles in these cells were mitochondria and smooth endoplasmic reticulum. Bannasch's (16) previous electron microscopic observations on hepatic nodules in rats produced similar conclusions. He observed clear, glycogen-containing cells, as well as lipid-storing cells. Basophilic cells contained increased rough endoplasmic reticulum. Eosinophilic hepatocytes were found to contain increased amounts of smooth endoplasmic reticulum. However, hepatocytes with increased numbers of mitochondria were also described. More recently, eosinophilic hepatocytes ("oncocytes") have been described in biopsies of patients with liver disease (17,18), as well as in human hepatocellular neoplasms (19). Eosinophilic hepatocellular neoplasms have previously been described in B6C3F1 mice treated with aromatic amines in hair dyes (20) and with nitrophen, a herbicide (21), as well as in C3H mice given phenobarbitone (22) and dieldrin (23).

Mallory bodies were observed only in dieldrin-fed animals, particularly in the eosinophilic neoplasms of the C57 and B6C3F1 strains (24). Elevation of enzyme activity was also characteristic of nodules in dieldrin animals (unpublished data). The relationship between dieldrin, Mallory bodies, elevated enzyme activity, cytoplasmic eosinophilia, and SER and mitochondrial proliferation in hepatic neoplasms of C57 and B6C3F1 mice requires further investigation.

Hepatocellular carcinomas were defined as tumors with a trabecular plate pattern and cytologic atypia. Although pulmonary metastases were observed infrequently in this experiment, they almost always occurred in tumors with this microscopic appearance. Hepatic nodules frequently contained benign areas, as well as areas showing a malignant pattern, as already defined. This finding certainly suggests that benign hepatic neoplasms may develop into hepatocellular carcinomas. Of interest in this respect is the marked difference in histochemical reactions between benign and malignant hepatocellular tumors. This suggests that transformation of benign into malignant hepatocellular tumors may be associated with a change in histochemical reaction pattern (25; also unpublished data).

REFERENCES

1. Reuber MD: Carcinomas and other lesions of the liver in mice ingesting organochlorine pesticides. Clin Toxicol 13:231-256, 1978.

2. Walker AIT, Thorpe E, Stevenson DE: The toxicology of dieldrin. Long term oral toxicity studies in mice. Food Cosmet Toxicol 11:415-432, 1973.

3. Grasso P, Hardy J: Strain differences in natural incidence and response to carcinogens. In: Mouse Hepatic Neoplasia, edited by WH Butler, PM Newberne, pp. 111-131. Amsterdam: Elsevier, 1975.

4. Lipsky MM, Tanner DC, Hinton DE, Trump BF: Reversibility, persistence, and progression of safrole-induced mouse liver lesions following cessation of exposure. Chap. 11 in this volume.

5. Kyriazis AP, Koka M, Vesselinovitch SD: Metastatic rate of liver tumors induced by diethylnitrosamine in mice. Cancer Res. 34:2881-2886, 1974.

6. Burstone M: Histochemical comparison of naphthol AS-phosphates for the demonstration of phosphatases. J Natl Cancer Inst 20:601-610, 1958.

7. Wachstein M, Meisel E: On the histochemical demonstration of glucose-6-phosphatase. J Histochem Cytochem 4:592, 1956.

8. Nachlas M, Tsou K, DeSouza E, Cheng C, Seligman A: Cytochemical demonstration of succinic dehydrogenase by the use of a new p-nitrophenyl substituted ditetrazole. J Histochem Cytochem 5:420-435, 1957.

9. Wachstein M, Meisel E: A comparative study of enzymatic staining reactions in the rat kidney with necrobiosis induced by ischemia and nephrotoxic agents (mercuhydrin and DL-serine). J Histochem Cytochem 5:204-208, 1957.

10. Vesselinovitch SD, Mihailovich N, Rao KN: Morphology and metastatic nature of nodular lesions in C57BL x C3HF1 mice. Cancer Res 38:2003-2010, 1978.

11. McDowell EM, Trump BF: Practical fixation techniques for light and electron microscopy. Comp Pathol Bull 9:1, 1977.

12. Spurr AR: A low-viscosity epoxy resin embedding medium for electron microscopy. J Ultrastruct Res 26:31-43, 1969.

13. Dragani TA, Sozzi G, Della Porta G: Differences in the susceptibility to the carcinogenic action of urethane in hybrid mice C57BL/6 x C3HF1 (B6C3) and C57BL/6 x BALB/C F1 (B6C). Tumori 67(2 suppl A):59, 1981.

14. Vesselinovitch SD, Mihailovich N: Kinetics of induction and growth of basophilic foci and development of hepatocellular carcinoma by diethylnitrosamine in the infant mouse. Chap. 5 in this volume.

15. Ruebner BH, Gershwin ME, Hsieh L, Dunn P: Ultrastructure of spontaneous neoplasms induced by diethylnitrosamine and dieldrin in the C3H mouse. J Environ Pathol Toxicol 4:237-254, 1980.

16. Bannasch P: Die Zytologie der Hepatocarcinogenese. Handbuch der Allg Path, vol. 6, pp. 123-276. Berlin: Springer Verlag, 1975.

17. Lefkowitch JH, Arborgh BAM, Scheuer PJ: Oxyphilic granular hepatocytes, mitochondrion rich liver cells in hepatic diseases. Am J Clin Pathol 74:432, 1980.

18. Gerber MA, Thung SN: Hepatic oncocytes. Incidence, staining characteristics and ultrastructural features. Am J Clin Pathol 75:498, 1981.

19. Wong LK, Link DP, Frey CF, Ruebner B, Tesluk H, Pimstone NR: Fibrolamellar hepatocarcinoma: Radiology, management and pathology. Amer J Radiol 139:172-175, 1982.

20. Reznik G, Ward JM: Carcinogenicity of the hair-dye component 2-nitro-p-phenylenediamine: Induction of eosinophilic hepatocellular neoplasms in female B6C3F1 mice. Food Cosmet Toxicol 17:493-500, 1979.

21. Hoover KL, Stinson SF, Ward JM: Histopathologic difference between liver tumors in untreated (C57BL/6 x C3H)F1 (B6C3F1) mice and nitrofen fed mice. Natl Cancer Inst 65:937-948, 1980.

22. Jones G, Butler WH: Morphology of spontaneous and induced neoplasia. In: Mouse Hepatic Neoplasia, edited by WH Butler, PM Newberne, pp. 21-57. Amsterdam: Elsevier, 1975.

23. Reuber MD: Histogenesis of hyperplasias and carcinomas of liver arising around central veins in mice ingesting chlorinated hydrocarbons. Pathol Microbiol 43:287-298, 1975.

24. Meierhenry EF, Ruebner BH, Gershwin ME, Hsieh LS, French SW: Mallory body formation in hepatic nodules of mice ingesting dieldrin. Lab Invest 44:392-396, 1981.

25. Ruebner B, Michas C, Kanayama R, Bannasch P: Sequential hepatic histologic and histochemical changes produced by diethylnitrosamine in rhesus monkey. J Natl Cancer Inst 54:1261-1267, 1976.

Chapter 10

# Characterization of Spontaneous and Chemically Induced Mouse Liver Tumors

**Frederick F. Becker**

INTRODUCTION

The studies described in this chapter had two major goals:

1. To associate a specific biologic behavior, i.e., benign or malignant, with a specific morphology of mouse liver tumors, and
2. To better understand the mechanisms that underlie the pathogenesis of spontaneous and chemically induced hepatocarcinogenesis.

However, several broad principles must be delineated to establish the frame of reference in which all such results can be interpreted. Some of these have been previously mentioned by others.

Although inbred mice are often used in studies of mouse hepatocarcinogenesis, inter- and intraexperimental variation may occur (1). Mice of a single inbred strain obtained from various sources may differ in many modalites, including their metabolic activation of carcinogens, immunologic responsivity, and others. In our hands (Stout and Becker, unpublished data), the liver carcinogen-activating fraction (microsomes $\pm$ cytosol) of C57Bl male mice obtained from three sources varied by more than a factor of 2 in their capacity to activate $\underline{N}$-2-fluorenylacetamide

---

The AFP values were determined by Dr. Stewart Sell, Department of Pathology, University of California, San Diego, under the auspices of National Cancer Institute grant CA 22227. Dr. James Bowen performed the examination for virus particles. Ann Sterling and Jean Burrell aided in the technical studies.

This investigation was supported by contract CP 65846 and grant CA 20659 awarded by the National Cancer Institute, Department of Health and Human Services.

(2-AAF) in a standard Ames assay. Further variation may occur even between mice of a single colony. Whether these findings result from genetic drift, "contamination" during breeding or other factors is as yet unknown. Chromosome and immunologic studies might help to explain these differences.

The "environment" of these experiments may also exert an extremely strong influence upon the results, particularly in the instance of spontaneous hepatocarcinogenesis. The number of mice per cage, presence or absence of carcinogens in the atmosphere, diet, ectoparasites, bedding, and even the texture of the food pellet have all been implicated as a cause of variations in spontaneous tumorigenesis (2-5). Extremes in calorie intake, protein, and fat content also exert an effect, but represent less unexpected and less easily identified factors (6).

These variations in milieu, except for those that are extreme, in our hands, have much less effect upon the chemically-induced liver tumors, as they will be described below. In summary, therefore, rigid consistency of the experimental conditions is required in such experiments. A sufficient number of control mice is a necessity, since one cannot extrapolate data from the controls of others.

In order to present this work economically, I will use only those references of others that present or establish broad major points.

MATERIALS AND METHODS

Male mice were obtained at 5 weeks of age and begun on their respective experimental regimens between 6 and 8 weeks of age. The strains used were C3H/HeN MTV⁻ (C3H), C57Bl/6N (C57Bl) and B6C3Fl (B6C3). All animals were obtained from the Division of Cancer Treatment, National Cancer Institute Animal Program. Although various diets and other environmental determinants influenced the incidence and time of onset of tumors and other results, as will be described, they did not affect the relationship between morphology and tumor biology.

The mice were kept six to a cage. The number was reduced by those demonstrated to have tumors as the experiments progressed. They were fed and given tap water _ad libitum_, save for a lesser number that received diethylnitrosamine (DEN) or phenobarbital (pb) in water. In most instances in which mice received chemical carcinogens, 2-AAF was mixed into the food and pelleted at a level of 0.03%. Chlordane (CRD) was fed at 25 ppm. No differences from the proposed classification occurred in mice that received 2-AAF at 0.045 or 0.06%, or CRD at 50 ppm. Smaller numbers of mice that received DEN or 3',4'-diaminoazobenzene conformed to the reported findings. Phenobarbital was administered at 0.05% in tap water and was freshly prepared 3 times per week.

On a monthly basis, the mice were observed clinically, and periorbital blood was obtained for determinations of alpha fetoprotein (AFP). In the majority of experiments (except where specified), the presence of tumors was detected by an elevation of circulating AFP, as has been described previously (7,8). A rise of 1 log above basal levels or a progressive elevation on 2 determinations was highly correlated (98% confidence) with the presence of a tumor. As a result of using this method, most of our tumors were detected relatively early in their course. These determinations were expertly performed by Dr. Steward Sell, Department of Pathology, University of California at San Diego.

A liver tumor was defined as a grossly evident lesion at least 1 mm in diameter. All tumors were obtained for histologic analysis and were given coded numbers not known to the author.

TRANSPLANTATION

The method of transplantation has been described previously (7,9,10). In brief, the tumor was rapidly minced by scalpel under sterile, cooled, 0.9% NaCl solution to fragments approximately 1-2 mm in diameter. Approximately 0.25 ml of fragments was subcutaneously injected into a shaved, sterile prepared area on the mid-right flank. In preliminary experiments, a similar intraperitoneal or intramuscular injection into the left thigh was performed. However, subcutaneous injection has proved to be almost as effective as any other type, and this location permits early detection and easy excision where required.

RESULTS

Morphology

A very large number of available reports are devoted to establishing a histologic criteria for diagnosing the lesions of mouse livers (2,11-23). These reports and many others have formed a valuable base for the understanding of the conditions under which each of the lesions and their many variations may appear. However, in an attempt to circumvent some of the terminologic pitfalls that have resulted from the many proposed systems, the classification that I have developed was derived post facto as a result of examining a series of functional characteristics of these tumors and, subsequently, by associating these with the histologic description, strain, and history (e.g., whether or not the mice had received carcinogen, phenobarbital). Therefore, all histologic descriptions are my own, performed on coded slides without identification of the mouse or treatment regimen. Although variations appeared, the classification as proposed fitted the majority of tumors and regimens.

Therefore, I propose a classification based on the term "mouse liver tumor" (MLT) type I, II, III, or IV (see Fig. 1 for detail). In brief, MLT types I and II were seen only in strains with a genetic propensity for spontaneous hepatotumorigenesis (C3H and B6C3( and conformed to previously described tumors: type A (17,19,23), type I (24), type I and II (2), and hepatic adenoma (12). Types I

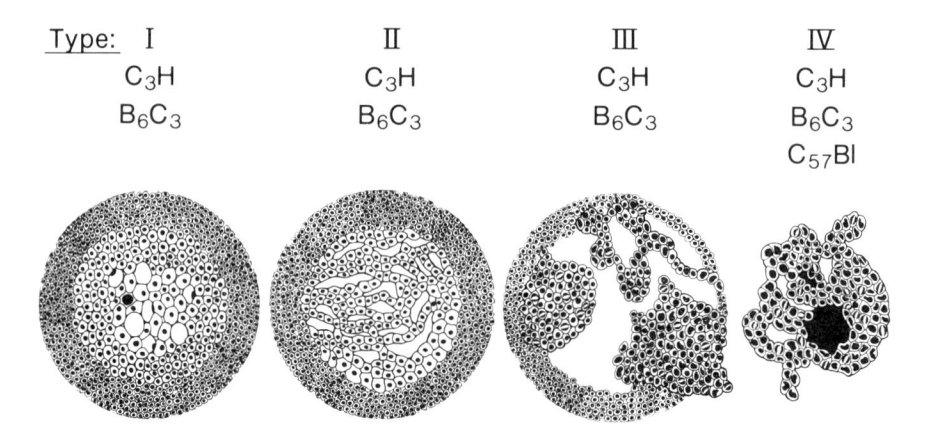

| Type: | I | II | III | IV |
|---|---|---|---|---|
| | C$_3$H | C$_3$H | C$_3$H | C$_3$H |
| | B$_6$C$_3$ | B$_6$C$_3$ | B$_6$C$_3$ | B$_6$C$_3$ |
| | | | | C$_{57}$Bl |

FIGURE 1. The typical appearance of the most common types of spontaneous and chemically induced mouse liver tumors. "Tumor" was defined as a grossly identifiable lesion 1 mm or greater in diameter. The strains of mice in which each type was detected are indicated. (Reproduced by permission Cancer Research 42(10):3918-3923, 1982)

The (idealized) zones of cell variants for MLT type I as depicted were seen mainly in larger tumors. Although the cytoplasm of the more central areas contained large amounts of glycogen or fat inclusions (some of which were external to cells), the nuclei were similar to those of the peripheral, proliferative zone.

MLT type II demonstrated trabeculation that ranged from less than 5% of the total section to 50% in some instances. Most of these foci were located central to the peripheral zone. These trabeculae were mainly composed of diploid cells with nuclei similar to those of the bulk of the tumor; they were mainly 1 to 2 cells thick; and, most strikingly, when trabeculae were traced to their termini, they ended in an intimate and normal-appearing relationship with the cells of the nontrabecular areas.

MLT type III were seen almost exclusively in mice exposed to known carcinogens. The characteristics of the foci were those associated with malignancy. Trabecular areas demonstrated high mitotic rates with occasionally aberrant forms; nuclear atypia of all degrees was observed; trabeculae were often many cells thick and had papillary projections. At the termination of these trabecular areas, they demonstrated invasion of the nodular parenchyma or often extended into the surrounding parenchyma. Cystic and necrotic areas were often present.

MLT type IV demonstrated the stigmata of frank malignancy with no visible remnant of types I and II. This was the sole tumor type seen in treated C57Bl mice.

and II were seen in these strains whether or not the mice had received exposure to a chemical agent.

The histologic appearance of type I tumors depended to some extent on their size. Tumors of more than 4 mm in diameter typically contained a peripheral zone composed of small, basophilic, diploid cells of approximately 2-8 cells in thickness. The borders of these cells were regular but poorly defined. The border between the tumor and normal hepatocytes was usually sharp and demonstrated some compression of the normal cells. The details of these findings with electron microscopy will be published in subsequent reports. Smaller tumors were composed almost entirely of these cells that seemed to represent the proliferative pool. Large tumors demonstrated a gradient of alteration toward the center of the tumor that resulted from the intracellular accumulation of various forms of inclusion materials, ranging from glycogen to fat; these have been well described in the literature.

Type II tumors were usually larger than 5 mm in diameter. They were characterized by zones or foci of trabeculation of various proportions. The trabeculation seen in type II tumors was composed mainly of cells that were cytologically similar to the cells of type I tumors, were 1-2 cells thick, and blended intimately into the cells of the nontrabecular areas. In some instances, this trabeculation was quite ornate and extensive. It is my belief that the majority of disputes in assigning a biologic or etiologic association with a given tumor morphology arise from decisions concerning the nature of these trabeculae. However, in my experience, when the trabeculae conformed to the above description - i.e., they possessed the cytologic aura of benignity - they were found most frequently in larger or "older" tumors in untreated mice.

The form of MLT designated type III is one that occasionally overlaps the appearance of type II tumors. Type III tumors appeared only in those mice with a genetic propensity for tumorigenesis that had received carcinogens. These tumors show, in varying degrees, remnants of MLT types I and II. However, either focally within them or replacing them to a considerable extent were foci of frank malignancy. In one form this also took a trabecular structure, thus leading to conflicts of decision with the type II tumor. However, in the main, the trabeculae of these foci bore characteristics that were clearly different from those of type II tumors. The cells usually had features of cytologic atypia that are indicative of their malignant nature. These features include an increased nuclear cytoplasmic ratio, nuclear abnormalities in regularity and chromatin pattern, and high mitotic rates. The trabeculae are often greatly increased in thickness, composed of as many as 5-10 cells or more. They form blunt papillary structures and demonstrate invasion of the nodule in which they arise. Thus, they do not have the intimate relationship with the other cell components of the tumor that the typical trabeculae of type II tumors have. However, it would be inaccurate to propose that difficulties in differentiating between the two types of tumors do not arise.

In addition to the trabecular form, the "nodules within nodules" of type III tumors may range from ill-defined areas with increased mitotic activity in cells of the type I tumor to nodular, anaplastic, cystic foci with the aberrant cytologic characteristics typical of cancers. These type III tumors may invade into their nodule of origin or outward into the normal parenchyma.

The tumors that have been designated MLT type IV are foci of frank malignancy that are seen in the parenchyma of liver. They may arise in strains with a genetic propensity to tumors or in mice that show little or no tendency toward tumorigenesis, both having received chemical carcinogens. These tumors conform to the descriptions in the literature of frank malignancy and rarely offer problems in diagnosis among pathologists. It is not the purpose of this presentation to discuss their site of origin, whether in benign proliferative lesions of the liver (25) or in preexisting tumors of types I and II (in those animals with a genetic tendency towards tumors).

Alpha Fetoprotein

In two previous papers from this laboratory, one that dealt with the appearance of spontaneous primary hepatocellular carcinomas (PHC) (8), and another with those tumors induced by chemical carcinogens (7), an extremely high correlation was demonstrated between the presence of tumor and an elevated circulating alpha fetoprotein (AFP) level. In spontaneous tumors arising in a subline of C3H mice and in tumors induced by 2-AAF and CRD in the C57Bl mouse, there was greater than 95% association with an elevated AFP value in the presence of a grossly visible tumor. The results of our currently reported studies with C3H, B6C3, and C57Bl mice confirmed these findings. Whether the tumors arose spontaneously in the livers of such mice or were induced by chemical agents, an identical association with circulating AFP was evident. No significant elevation of circulating AFP was detected unless a grossly visible tumor was evident. This, then, is quite different from the situation in which an extremely early elevation of AFP may occur in rats given several carcinogens (26,27). No significant differences in the levels of AFP were detected regardless of cell type of the PHC.

Transplantation

Two of the characteristics of our transplantation experiments are crucial to their interpretation. The transplantation success rate for any given tumor type (I-IV) has remained approximately constant over $4\frac{1}{2}$ years. In addition, observation of smaller groups of recipient mice indicated that detection of successful transplants that appeared at later than one year of observation added less than 7% to the success rate of any tumor type. Therefore, observation for 1 yr after transplantation was considered appropriate.

# CHARACTERIZATION

No significant differences in success rate were determined between types I and II or III and IV in a given strain, and therefore, to increase the numbers involved, these will be grouped.

Of all group I and II transplants, 36% were successful in C3H mice (121/336). The range of time until definitive detection of tumor (usually at a diameter of approximately 1 cm) was 6-11 months, with a mean of 7½ months. Of transplanted tumors of types II and IV, 64% were successful in C3H mice (52/81). The range of detection was 3-9 months, with a mean of 4.5 months.

Of tumors of type IV, 74% were successfully transplanted in C57Bl mice. The range of time from transplant until appearance was 1.5 to 8 months, with a mean of 3.5 months.

Although only a small number of tumors were followed with repeated transplantation, one additional finding related to the tumor histotype warrants mention. Six to eight type I tumors sustained their original histologic pattern for between two and four transplantation generations. Two of eight demonstrated an increasingly trabecular pattern, one achieving a fully papillary and trabecular pattern after three passages. In general, types III and IV tumors became increasingly homogeneous with transplantation, selecting for the more atypical cell types.

## VIRUS STUDIES

Tumors that arose spontaneously in C3H and B6C3 mice, as well as those that were induced by exposure to chemical carcinogens in these and C57Bl mice, were examined by electron microscopy for the presence of viral particles. Eight of eight spontaneous tumors in C3H mice demonstrated type C virus production with budding from membranes into extracellular spaces, most prominently into canaliculi. Particles showing typical features of budding type C viruses could be seen along the plasma membrane of many sections.

Six PHCs (4 type III, 2 type IV), induced in C3H mice by chemical agents (DEN, 2; 2-AAF, 3; CRD, 1), all demonstrated low levels of viral particles of identical type. The electron microscopic examination of the cells of these tumors also confirmed the impression of their lesser degree of differentiation obtained by histologic analysis when compared to cells of tumor types I and II.

No viral particles could be discerned on examination of multiple sections from four of five B6C3 spontaneous tumors. After examination of four areas of a fifth tumor of this type, a single, abortive viral particle was observed.

No viral particles were observed in nine chemically induced (type IV) tumors in C57Bl mice (2-AAF, 5; CRD, 4).

Phenobarbital Exposure

To determine the response of these strains to a putative tumor promoting agent, 16 male mice of each strain were begun on 0.05% phenobarbital (pb) in their drinking water at 6-8 weeks of age (Table 1). An additional 16 mice received untreated water. All survived until 1 year of treatment, at which time the C3H and B6C3 mice were sacrificed. The control mice in both C3H and B6C3 groups had relatively high tumor incidences of 61 and 29%, respectively. However, those mice that received phenobarbital not only demonstrated 100% tumor incidence in each of these strains, but also showed an increase in tumor number and size. Indeed, every treated C3H mouse had at least one lobe totally "converted" to PHC, as did two B6C3 mice.

Histologic examination indicated that all of the PHCs in treated and untreated mice were either type I or type II. Although a suggestion of a higher percentage of type II was seen in pb-treated mice, this was not statistically significant.

TABLE 1. Effect of Phenobarbital on Incidence of Primary Hepatic Carcinomas[a]

| Treatment | PHC[b] | Number/mouse[c] | Largest PHC |
|---|---|---|---|
| C3H | | | |
| -pb | 61% | 1-4 | 1 cm |
| +pb | 100% | 10 | Total conversion,[d] 22/16 mice |
| B6C3 | | | |
| -pb | 29% | 1-2 | 0.7 cm |
| +pb | 100% | 3-8 | Total conversion,[d] 2/16 mice |
| C57Bl | | | |
| -pb | 0% | 0 | |
| +pb | 0% | 0 | |

[a] Sixteen male mice of each strain received 0.05% pb in their drinking water; 16 additional males of each strain received untreated water. Incidence of PHC was determined following sacrifice at the times indicated.
[b] Incidence of primary hepatic carcinoma when sacrificed at 1 year of exposure (C3H and B6C3) or at 18 months (C57Bl).
[c] Number of PHC/tumor-bearing mouse (range).
[d] Number of totally converted lobes.
(Reproduced by permission Cancer Research 42(10):3918-3923, 1982)

Since the C57Bl mice demonstrated no alterations of AFP levels at 1 year, they were maintained on pb for a total of 18 months. Neither treated nor untreated C57Bl mice demonstrated any tumors when killed at that time.

Effect of Carcinogens on MLT Types I and II

A study was performed to examine the impression that the carcinogens used in these experiments increased the incidence of tumors of types I and II as well as induced types III and IV (Table 2). Groups of C3H mice received basal diet, basal diet plus 0.05% pb in their water, 0.03% 2-AAF, or 25 ppm CRD. Eight mice from each treatment group were sacrificed at the designated times. Table 2 demonstrates the results. At each time studied, the incidence of MLT types I and II was greater in groups of C3H mice receiving pb, 2-AAF, or CRD. The incidence of tumor types I and II appeared to be independent of the induction of types III and IV. Although the enhancement of types I and II was evident, insufficient animals were available to establish the significance of this.

Influence of Diets

One interesting finding that was somewhat incidental to the main thrust of the study was the striking influence of various "standard" diets upon the incidence of type I and type II tumors. A number of authors have previously referred to this phenomenon (28). A significant proportion of our experiments have been performed with a semipurified diet no. 101 (Bio-Serve Corp., Frenchtown, N.J.).

TABLE 2. Effect of Carcinogens on the Incidence of MLT Types I and II[a]

| Treatment | Age (weeks) 26 | 39 | 52 |
|---|---|---|---|
| Basal diet | 0/8[b] | 1/8 (12.5%) | 5/8 (62.5%) |
| pb[c] | 1/8 (12.5%) | 3/8 (37.5%) | 8/8 (100%) |
| AAF[d] | 1/8 (12.5%) | 3/8 (37.5%) | 7/8 (87.5%) |
| CRD[e] | 2/8 (25.0%) | 4/8 (50.0%) | 8/8 (100%) |

[a] C3H mice received basal diet, 0.05% pb, 0.03% 2-AAF, or 25 ppm CRD. Eight mice of each group were sacrificed at the times indicated, and the incidence of MLT type I or II was determined.
[b] Incidence of primary hepatic carcinogens.
[c] 0.05% pb in drinking water administered continuously.
[d] 0.03% 2-AAF in basal diet fed continuously.
[e] 25 ppm CRD in basal diet fed continuously.
(Reproduced by permission Cancer Research 42(10):3918-3923, 1982)

The purpose of using this diet was to avoid a number of the components common to standard laboratory diets. These include antioxidants, variable quantities of vegetable matter, hormones, etc., all of which can influence carcinogenesis. Intriguingly, a wide variation of effects was noted among standard commercial diets. As is evident from Table 3, the Agway diet (Charles River Farms) caused a significant delay in the appearance of MLT types I and II in C3H mice when compared with results reported in the literature and with our own findings with other regimens. The delay was less significant if Agway was begun after an initial period of exposure to the no. 101 diet. The tumor incidences in mice achieved using Wayne Lab Blox (Allied Mills, Inc., Chicago, IL.) were similar to those reported in the literature with this strain, while mice fed no. 101 demonstrated a significantly earlier appearance of PHC.

DISCUSSION

The importance of an acceptable classification of mouse liver tumors cannot be underestimated. The response of the mouse liver to chemical agents represents the foundation of many bioassay programs. In addition, the availability of strains with a genetic predisposition to hepatotumorigenesis makes the mouse a particularly

TABLE 3. Incidence of MLT Tumor Types I and II in C3H Mice as a Function of Diet[a]

| Diet | Age (weeks) | | | |
| --- | --- | --- | --- | --- |
|  | 39 | 52 | 65 | 78 |
| No. 101[b] | 5-12% | 41-62% | 67-81% | 84-98% |
| Agway | 0 | 0 | ---[c] | 41 |
| No. 101 + Agway (10 weeks)[d] | 0 | 32 | --- | 74 |
| No. 101 + Agway (20 weeks)[d] | 0 | 51 | --- | 90 |
| Wayne Lab Blox | 0 | 38 | --- | 84 |

[a] At 6-8 weeks of age (except where indicated), C3H mice were begun on diet no. 101, Agway, or Wayne Lab Blox. The presence of MLT was determined by circulating levels of AFP, as described in the text.
[b] The incidences with no. 101 represent an aggregate of 1350 mice. More than 20 individual groups are represented in this aggregate. All other dietary regimens had 16 mice each.
[c] Not performed.
[d] These mice were fed diet no. 101 for 10 or 20 weeks prior to initiation of the Agway ration.

excellent model in which to study factors that influence carcinogenesis. Therefore, the almost endless debates in meetings and in published works that argue the nature of the mouse liver tumors have inhibited progress using this valuable species.

Two of the major problems in formulating an acceptable classification of mouse liver tumors have been the failure to give sufficient recognition to the significance of tumors of spontaneous origin and the attempt to limit too greatly the classes of tumor, i.e., benign or malignant. Further, discussions of the classification have often been dominated by emphasis on a lesser number of lesions about which there is controversy, rather than on the broad areas of agreement. Thus, if one examines many of the proposed systems, there is general agreement that MLT type I (hyperplasia; type A, type I; adenoma; etc.) are at most, minimally malignant in terms of metastasis and overt invasion. This has been expecially well demonstrated by the studies of Vesselinovitch in numerous studies with many agents (13,18). These lesions are seen solely in mice with a genetic predisposition towards hepatocarcinogenesis, whether or not they have been exposed to chemical agents. Also, if one is reasonably strict in limiting the definition of MLT type II to those tumors in which the trabecular cytology is composed of benign-appearing cells of 1-3 cells thickness, MLT type II also demonstrated the functional characteristics of a relatively low-grade malignancy. Thus, in all mouse strains tested, only two of the four tumor types (III and IV) were conclusively associated with exposure to and *de novo* induction by carcinogens. By this, I imply a holocarcinogen induction.

What then of the problem of interpretation of increased incidence of MLT types I and II with exposure to chemical agents? Is this evidence of carcinogenicity? The current study strongly supports the proposal of Peraino et al. (29) that phenobarbital acts only as a promoting agent. Although phenobarbital exposure enormously increased the number, incidence and total mass of PHC, it did so only in the two strains of mice with genetic predisposition to spontaneous tumors. This has also been suggested in other published studies (17,30,31). Further, the increase in PHC induced by phenobarbital in the current study was mainly limited to MLT types I and II, a finding also reported by Peraino et al. (29). Phenobarbital did not induce tumors of any type in C57Bl mice, a strain that, in our laboratory, does not demonstrate a tendency toward spontaneous hepatocarcinogenesis. This indicates that phenobarbital requires some form of genetic initiation if it is to act as a promoter, a useful and intriguing finding. It is also interesting that C57Bl mice have no genetic background with C3H mice, in contrast to the majority of strains used in bioassays (32).

Although this study was only performed as a pilot, it appeard that known carcinogens (in addition to inducing MLT types III and IV) may also cause an increase in MLT types I and II in appropriate strains. This has been previously reported by Gellatly (2) and suggested by others. Together with the results in pb-exposed mice, this finding indicates that increases in MLT types I and II in such strains, during exposure to chemical, may be the result of a promotional capability or component of the agent, rather than holocarcinogenesis.

In summary, I would like to stress several points in regard to the proposed classification:

1. This classification can be considered an extension of others and in particular that of Gellatly, who recognized the need for expansion of previous systems.
2. It is not suggested that it is definitive of all tumors found in mouse livers, since it ignores vascular, stromal, and ductular elements, as well as the hepatoblastoma (12,23,24).
3. It does not attempt to include all variations of parenchymal lesions, nor do I assume that it will eliminate all disputes regarding these, especially regarding differentiating MLT types II and III. Several descriptions of MLT type II tumors (in untreated older mice with very large tumors) are similar in appearance to type III tumors (12,19,20,33).
4. It does not imply any conclusions concerning the possibility that MLT types I and II are especially susceptible to progression to MLT types III and IV under the influence of carcinogens. To do this would require an extraordinary statistical analysis of the relative numbers of each tumor type in several strains at each time point.

The classification offered herein attempts to emphasize those lesions and their component structures upon which agreement is most likely. Within this working definition, it is proposed that the classification is a useful one in terms of applicability, since the types of tumors described appear to include those that comprise the major proportion in most studies. Our failure to distinguish tumors of spontaneous origin from those induced by chemical agents by their production of AFP, viruses, or transplantability reinforced our dependence upon histologic criteria.

As a result of those studies that suggest a promotional capability for phenobarbital and for carcinogens (in terms of MLT types I and II), it is also important to stress that, in bioassays or in experiments in which mice with a genetic predisposition are examined, histologic identification of these tumors that are specifically associated with carcinogen exposure (MLT types III and IV) must be utilized to define the carcinogenicity of the agent.

REFERENCES

1. Tarone RE, Chu KC, Ward, JM: Variability in the rates of some common, naturally-occurring tumors in Fischer-344 rats and (C57Bl/6N x C3H/HeN) $F_1$ (B6C3$F_1$) mice. J Natl Cancer Inst 66:1175-1181, 1981.

2. Gellatly JBM: The natural history of hepatic parenchymal nodule formation in a colony of C57Bl mice with reference to the effect of diet. In: Mouse Hepatic Neoplasia, edited by

WH Butler and PM Newbern, pp. 77-109. Amsterdam: Elsevier, 1974.

3. Mizutani T, Mitsuoka T: Effect of intestinal bacteria on incidence of liver tumors in gnotobiotic C3H/H3 male mice. J Natl Cancer Inst 63:1365-1370, 1979.

4. Schoental R: Variation in the incidence of "spontaneous" tumors. Br J Cancer 39:101, 1978.

5. Vlahakis G: Possible carcinogenic effects of cedar shavings in bedding of C3H-A$^{vy}$ fB mice. J Natl Cancer Inst 58:149-150, 1977.

6. Tannenbaum A, Silverston H: The influence of the degree of caloric restriction on the formation of skin tumors and hepatomas in mice. Cancer Res 9:724-727, 1949.

7. Becker FF, Sell S: α-Fetoprotein levels and hepatic alterations during chemical carcinogenesis in C57BL/6N mice. Cancer Res 39:3491-3494, 1979.

8. Becker FF, Stillman D, Sell S: Serum α-fetoprotein in a mouse strain (C3H-A$^{vy}$ fB) with spontaneous hepatocellular carcinomas. Cancer Res 37:870-872, 1977.

9. Lapeyre JN, Walker MS, Becker FF: DNA methylation and methylase levels in normal and malignant mouse hepatic tissues. Carcinogenesis 2:873-878, 1981.

10. Becker FF: Inhibition of spontaneous hepatocarcinogenesis in C3H/HeN mice by transplanted hepatocellular carcinomas. Cancer Res 41:3320-3323, 1981.

11. Andervont HB, Bunn TB: Transplantation of spontaneous and induced hepatomas in inbred mice. J Natl Cancer Inst 13:455-503, 1952.

12. Frith CH, Ward JM: A morphologic classification of proliferative and neoplastic hepatic lesions in mice. J Environ Pathol Toxicol 3:329-351, 1980.

13. Kyriazis AP, Koka M, Vesselinovitch SD: Metastatic rate of liver tumors induced by diethylnitrosamine in mice. Cancer Res 34:2881-2886, 1974.

14. Reuber MD: Histopathology of carcinomas of the liver in mice ingesting dieldrin or aldrin. Tumori 62:463-472, 1976.

15. Reuber MD, Ward JM: Histopathology of liver carcinomas in (C57BL/6N X C3H/HeN)F$_1$ mice ingesting chlordane. J Natl Cancer Inst 63:89-91, 1979.

16. Reubner BH, Gershwin ME, Hsieh L, Dunn P: Ultrastructure of spontaneous hepatic neoplasms and of neoplasms induced by

diethylnitrosamine or dieldrin in the C3H mouse. Fed Proc 38:1451, 1979.

17. Thorpe E, Waler AIT: The toxicology of dieldrin (HEOD*). II. Comparative long-term oral toxicity studies in mice with dieldrin, DDT, phenobarbitone, β-BHC and γ-BHC. Food Cosmet Toxicol 11:433-442, 1973.

18. Vesselinovitch SD, Mihailovich N, Rao KVN: Morphology and metastatic nature of induced hepatic nodular lesions in C57BL x C3H $F_1$ mice. Cancer Res 38:2003-2010, 1978.

19. Walker, AIT, Thorpe E, Stevenson DE: The toxicology of dieldrin (HEOD*). I. Long-term oral toxicity studies in mice. Food Cosmet Toxicol 11:415-532, 1972.

20. Ward JM: Morphology of hepatocellular neoplasms in B3C3F1 mice. Cancer Lett 9:319-325, 1980.

21. Ward JM, Griesemer RA, Weisburger JH: The mouse liver tumor as an endpoint in charcinogenesis tests. Toxicol Appl Pharmacol 51:389-397, 1979.

22. Ward JM, Vlahakis G: Evaluation of hepatocellular neoplasms in mice. J Natl Cancer Inst 61:807-811, 1978.

23. Jones G, Butler WH: Morphology of spontaneous and induced neoplasia. In: Mouse Hepatic Neoplasia, edited by WH Butler, PM Newberne, pp. 21-59. Amsterdam: Elsevier, 1974.

24. Engelese LD, Hollander CF, Misdorp W: A sex-dependent difference in the type of tumors induced by dimethylnitrosamine in the livers of C3Hf mice. Eur J Cancer 10:129-135, 1974.

25. Grasso P, Hardy J: Strain difference in natural incidence and response to carcinogens. In: Mouse Hepatic Neoplasia, edited by WH Butler, PM Newberne, pp. 111-132. Amsterdam: Elsevier, 1974.

26. Becker FF, Sell S: Early elevation of alpha 1-fetoprotein in N-2-fluorenylacetamide hepatocarcinogenesis. Cancer Res 34:2489-2494, 1974.

27. Becker FF, Horland AA, Shurgin A, Sell S: A study of alpha 1-fetoprotein levels during exposure to 3'-methyl-4-dimethylaminoazobenzene and its analogs. Cancer Res 35:1510-1513, 1975.

28. Silverstone H: The effect of rice diets on the formation of induced and spontaneous hepatomas in mice. Cancer Res 8:309-317, 1948.

29. Peraino C, Fry RJM, Staffeldt E: Enhancement of spontaneous hepatic tumorigenesis in C3H mice by dietary phenobarbital. J Natl Cancer Inst 51:1349-1350, 1973.

30. Butler WH, Hempsall V: Histochemical observations on nodules induced in the mouse liver by phenobarbitone. J Pathol 125:155-161, 1978.

31. Ponomarkov V, Tomatis L, Turusov V: The effect of long-term administration of phenobarbitone in CF-1 mice. Cancer Lett 1:165-172, 1976.

32. Staats J: Standardized nomenclature for inbred strains of mice: Seventh listing. Cancer Res 40:2083-2128, 1980.

33. Hoover KL, Ward JM, Stinson SF: Histopathologic differences between liver tumors in untreated (C57Bl/6 X C3H)$F_1$ (B6C3$F_1$) mice and nitrofen-fed mice. J Natl Cancer Inst 65:937-948, 1980.

Chapter 11

# Reversibility, Persistence, and Progression of Safrole-Induced Mouse Liver Lesions following Cessation of Exposure

**Michael M. Lipsky, David C. Tanner, David E. Hinton, and Benjamin F. Trump**

INTRODUCTION

Studies in our laboratory have been directed toward the identification and characterization of potential preneoplastic and neoplastic alterations in BALB/c mouse livers induced by dietary safrole (4-allyl-1,2-methylenedioxybenzene) exposure (1-4). Previous investigators described a variety of safrole-induced toxic and neoplastic hepatic lesions in rats and mice (5-8). We focused our attention on areas of centrolobular hepatocytomegaly, focal lesions of phenotypically altered cell populations (foci), hepatocellular adenomas (HA) and hepatocellular carcinomas (HPC). The morphological appearance, enzyme histochemical pattern, and other features of these lesions were described in previous publications (1-4). The HPC were shown to metastasize and were readily transplantable, demonstrating their neoplastic nature. Controversy still exists, however, over the neoplastic potential of other liver lesions, especially the foci and HA.

A few investigators have studied the reversibility/irreversibility of nonmalignant nodules (HA) induced by carcinogens in mouse and rat liver (9,10). Their results demonstrated variability in response and an inability to identify, using known cell markers, the lesions that reverse (revert to a normal phenotype) from those that persist or progress. Our studies were designed to quantitatively assess the reversibility, persistence, and/or progression of specific, safrole-induced liver lesions after cessation of safrole treatment. This was accomplished by morphometric analysis of liver sections for foci of cellular alteration and the counting and measuring of grossly visible HA and HPC.

---

This work was supported by a contract from the National Cancer Institute, No-1-CP-65752.

MATERIALS AND METHODS

The experimental design is diagrammed in Table 1. Young (6-8 week old) male BALB/c mice (Charles River, Wilmington, MA) were randomly housed 4-5 per plastic cage on a 12-hour light/dark cycle. Food and water were available ad libitum. The food was a commercial laboratory chow (Purina) for the control mice, or the laboratory chow to which safrole (4000 ppm) was added (Bio-Serve, Frenchtown, N.J.). At the appropriate time intervals, 15-25 mice from each group were killed according to the protocol in Table 1. The protocol is divided into four sets of studies. Groups 1, 4, and 6 were used to evaluate the effects of 16, 36, and 52 weeks of continuous safrole feeding. Groups 1, 2, and 3 were compared to assess the effects of 16 weeks of safrole exposure, and 16 weeks of safrole followed by 36 and 52 additional weeks of control feeding (i.e., absence of the carcinogen). Groups 4 and 5 compared 36 weeks of continuous safrole feeding and 36 weeks safrole plus 36 weeks additional control feeding. Groups 6 and 7 compared a 52-week exposure to safrole plus 36 additional weeks on control diet. Group 8 served as a negative control group for all experiments. All groups were from the same initial set of mice, age-matched and simultaneously fed. This design was adopted to evaluate the behavior, nature, and autonomous growth potential of hepatocytic lesions induced by safrole for the various time intervals in the protocol.

TABLE 1. Experimental Design

| Groups | Weeks |
| --- | --- |
| | 8   16   24   36   52   68   76   84   92 |
| 1 | XXXXXXXXXXXS |
| 2 | XXXXXXXXXXXX_____S |
| 3 | XXXXXXXXXXXX_____S |
| 4 | XXXXXXXXXXXXXXXXXXXXXXS |
| 5 | XXXXXXXXXXXXXXXXXXXXXX_____S |
| 6 | XXXXXXXXXXXXXXXXXXXXXXXXXXXS |
| 7 | XXXXXXXXXXXXXXXXXXXXXXXXXXX_____S |
| 8 | S    S    S    S    S         S          S |

XXXX = Safrole treatment.
____ = Control diet feeding.
S = Sampling point.
15-20 mice were sampled at each S.
10 Control mice were sampled at each point, except at 92 weeks (18).

The livers were examined, and all grossly visible lesions were counted and measured. A section from the center of each lesion was fixed in 4% formaldehyde for subsequent histological diagnosis. Nonnodular livers were sliced into 2-3-mm thick sections, and random sections from all lobes were either frozen in liquid nitrogen for enzyme histochemical analyses or fixed in 4% formaldehyde. Three to five slices from each lobe were routinely sampled for enzyme histochemistry. Formaldehyde-fixed tissue was embedded in paraffin, sectioned, and stained with hematoxylin and eosin (H&E). Fresh-frozen tissue was cut at 8-10 μm in a cryostat and histochemically stained for 3 enzymes. Gamma-glutamyl transpeptidase (GGT) was determined by the method of Ruttenberg et al. (11). Glucose-6-phosphatase (G-6-pase) was determined by the method of Wachstein and Meisel (12). Glucose-6-phosphate dehydrogenase (G-6-PDH) was determined by the method of Negi and Stephen (13).

Morphometric analysis was performed using the standard area-perimeter and stereology programs in a Zeiss Videoplan system. The equations in the programs are identical to those described by Underwood (14). The parameters reported herein include: $V_v$, a ratio of the total volume of lesions to the total test volume; $N_v$, the number of lesions/mm$^3$; and $V_v$, the mean volume (mm$^3$) of the individual lesions.

RESULTS

Morphological Alterations

In order to place the morphometric data in perspective, we will briefly describe the morphology of the important lesions seen in our previous studies. A panlobar, centrolobular hepatocytomegaly developed in the safrole-treated mice after 2-4 weeks of feeding. Affected hepatocytes were round and stained acidophilic in H&E sections. When examined by electron microscopy, these cells contained large areas of smooth endoplasmic reticulum. The activity of G-6-pase was decreased when compared to the activity in livers from control mice. However, GGT activity in hypertrophied hepatocytes was unchanged when compared to activity in normal hepatocytes. Foci of phenotypically altered hepatocytes appeared after 24 weeks of safrole treatment. In H&E stained sections, they were composed of basophilic, acidophilic, or clear cells. A larger number of foci was visualized by altered staining of marker enzymes, including decreased activity of G-6-pase and/or increased activity of GGT and G-6-PDH (Figs. 1 and 2). There was a large degree of variability in reaction patterns of individual foci when stained for these 3 enzymes. HA were noted as early as 36 weeks after the initiation of safrole feeding. They caused compression of adjacent parenchyma and were composed of cells with a heterogeneous staining pattern in H&E stained sections (Fig. 3). Cells in HA generally contained decreased G-6-pase and increased GGT and G-6-PDH activity in histochemically stained sections (Fig. 4). These lesions did not produce metastases and were identical to the nonmalignant nodules called a variety of names, including hyperplastic nodule and neoplastic nodule, by different investigators. The HPC were

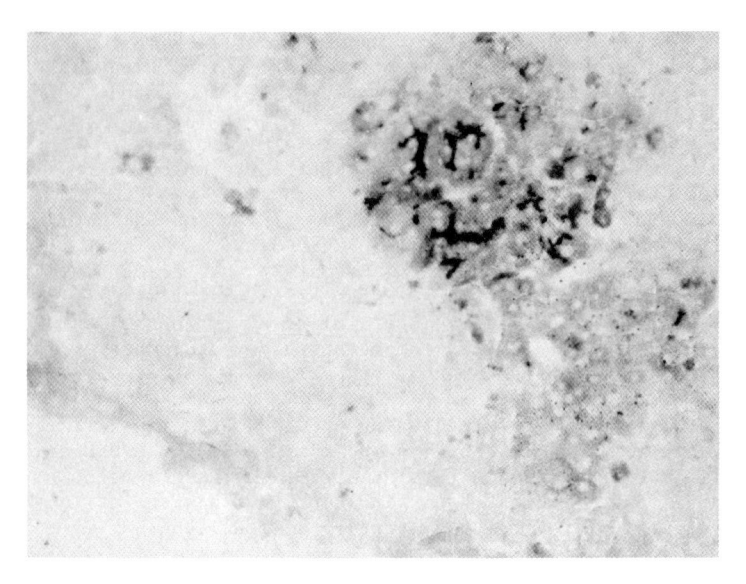

FIGURE 1. Focus of GGT-positive hepatocytes after 36 weeks of safrole feeding, X75. From Lipsky et al. (3).

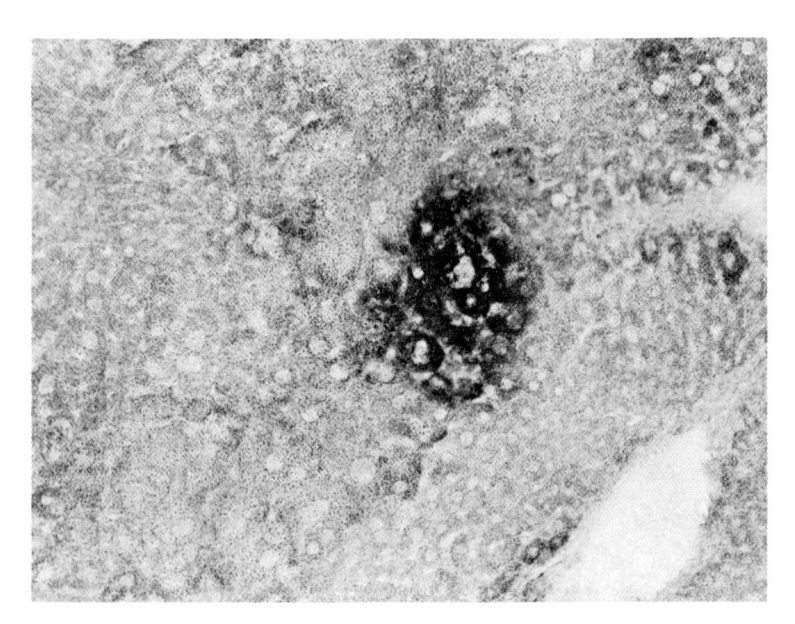

FIGURE 2. Small focus of G-6-PDH-positive hepatocytes, X100. From Lipsky et al. (3).

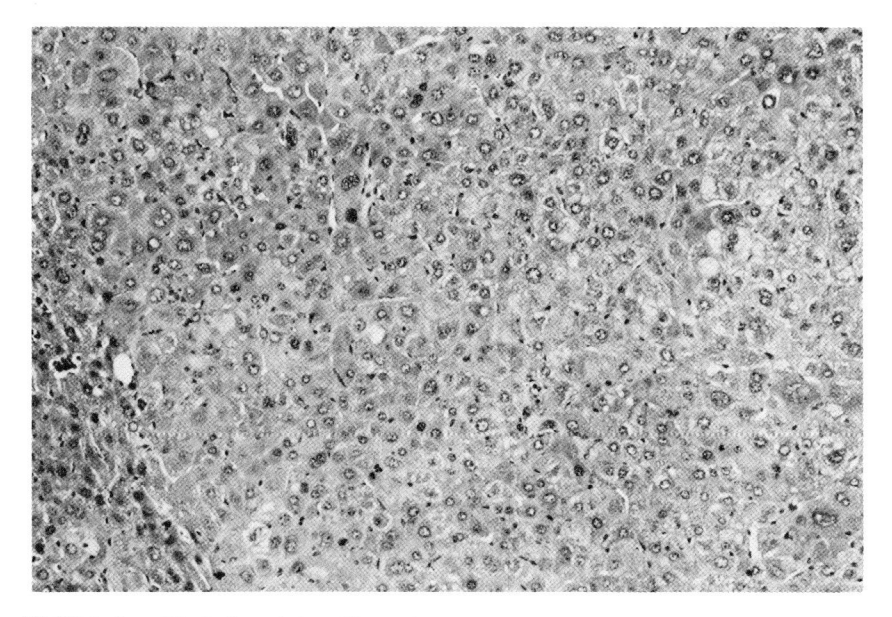

FIGURE 3. HA induced by 52 weeks of safrole feeding, X50.

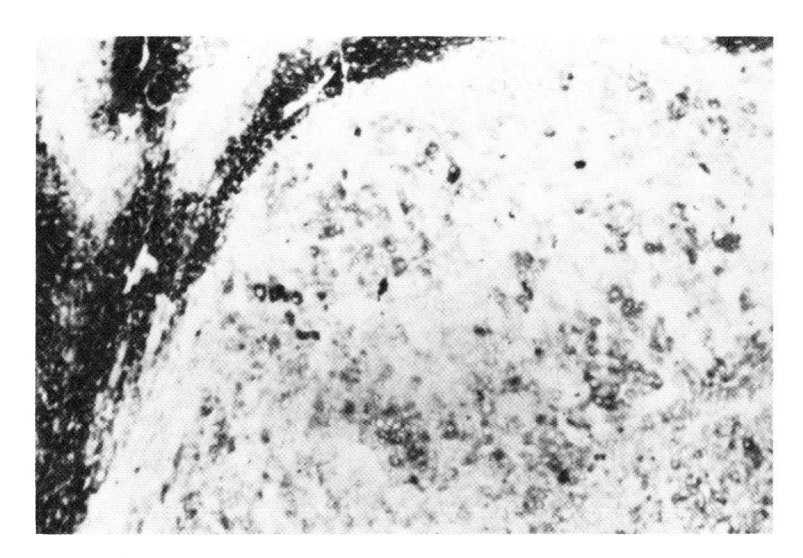

FIGURE 4. Decreased G-6-pase activity in a HA, X50.

composed of a heterogeneous cell population arranged in trabecular and pseudoglandular patterns (Fig. 5). Enzyme histochemical staining varied between HPC as well as between cells in the same HPC. In general, however, the enzyme patterns were similar in the HPC to those seen in the HA. The HPC metastasized and were readily transplantable.

Morphometric Analysis of Foci

Our previous work had shown that the enzyme markers were more effective than H&E staining in localizing focal populations of altered hepatocytes (2,3). Since the patterns were variable between foci, the initial quantitative experiments evaluated the phenotypic distribution of the 3 enzymes in mice continuously fed safrole for 36 and 52 weeks. Table 2 summarizes the 36-week data. Over 86% of the total population of foci identified by the enzyme alterations contained GGT. Only 1.2% of the foci displayed altered reactions for all 3 enzymes. The results seen after 52 weeks of safrole feeding were similar to those at 36 weeks. The control mouse livers did not contain enzyme-altered foci at any time period examined. Based on these results, we elected to utilize only GGT as a marker for foci in the morphometric analysis. This enabled us to efficiently count and measure the majority of foci present in the livers of treated mice. Table 3 summarizes the morphometric data. These are subdivided into related groups and presented in graphic form in Fig. 6-8. The results of the continuous exposure groups are presented in Fig. 6. There were no GGT-positive foci before 16 weeks. Only 1 of 10 mice had liver foci at 16 weeks so no data are depicted in the graph. By 36 weeks of safrole feeding, the relative volume of liver occupied by GGT-positive foci (Vv) was $0.011 \pm .0004$. This value did not change appreciably after 52 weeks of continuous safrole feeding ($0.012 \pm .003$). However, the number of foci/mm$^3$ of liver (Nv) declined from 1.08 to 0.31 between 36 and 52 weeks. This was accompanied by an increase in the mean volume (Vq) of individual foci from 0.0013 to 0.136 mm$^3$. These data show maximum foci development between 24 and 36 weeks, with a steadystate existing at 36-52 weeks.

Figure 7 presents the quantitative data from the 3 groups of mice fed safrole continuously for 16 weeks. This feeding schedule produces a panlobar, centrolobular hepatocytomegaly but no foci, HA, or HPC. GGT was not present in stainable amounts in hepatocytes. However, when the mice were removed from the safrole diet and fed the control diet for an additional 36 or 52 weeks, foci, HA, and HPC developed. The Vv of foci was similar after 36 and 52 additional weeks of control diet. The Nv decreased from $2.21 \pm 0.42$ to $0.21 \pm 0.05$ mm$^3$ over this time. As with the continuous feeding groups reported above (36 or 52 weeks of safrole), the Vq increased from 0.013 after 36 weeks of control feeding to 0.062 after 52 weeks of control feeding. These data demonstrated a decrease in foci number but an increase in volume of individual foci, while the total volume of GGT-positive lesions per unit volume of liver remained the same.

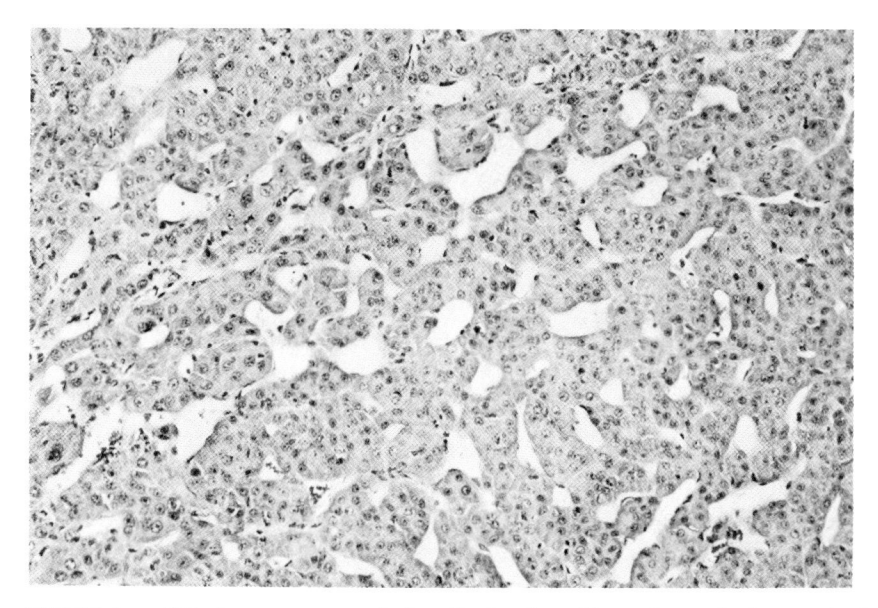

FIGURE 5. HPC from a mouse fed safrole for 36 weeks and then switched to the control diet for an additional 36 weeks, X50.

TABLE 2. Distribution of Hepatocellular Foci by Enzymatic Phenotype

| | | |
|---|---|---|
| 1. | GGT[a] | 77.4% |
| 2. | G-6-pase | 8.7 |
| 3. | G-6-PDH | 4.9 |
| 4. | GGT+G-6-pase | 3.8 |
| 5. | GGT+G-6-PDH | 4.0 |
| 6. | GGT+G-6-pase+G-6-PDH | 1.2 |
| 7. | G-6-pase+G6PDH | 0% |

[a] % Foci marked by GGT = 86.4% (represents all sections sampled from 10 mice after 36 weeks of safrole treatment).

TABLE 3. Stereology of Hepatocellular Foci

| Weeks on protocol | Vv | Nv | Vq |
| --- | --- | --- | --- |
| 16 S | - | - | - |
| 16 S+36 cont. | .0043 ± .0009 | 2.21 ± .416 | .013 ± .009 |
| 16 S+52 cont. | .0044 ± .0011 | 0.212 ± .054 | .062 ± .021 |
| 36 S | .011 ± .0004 | 1.077 ± .196 | .0013 ± .0002 |
| 36 S+36 cont. | .031 ± .0024 | .113 ± .017 | .025 ± .022 |
| 52 S | .012 ± .003 | .312 ± .081 | .136 ± .055 |
| Control | 0 | 0 | 0 |

[a]Analysis based on 7-10 sections (6-7 μm) from 5-7 animals stained for GGT. The results are expressed as mean values ± standard error of the mean.

S = Safrole feeding; cont. = control diet feeding.

Vv = Volume of lesions in unit volume of structure; a ratio of volume of lesions to test or reference volume ($mm^3/mm^3$).

Nv = Number of lesions per unit test volume ($\#/mm^3$).

Vq = Mean volume of lesions ($mm^3$).

Figure 8 presents the data from the groups that received 36 weeks of continuous safrole feeding. In this case, the Vv increased from 0.011 to 0.031 when mice exposed to safrole for 36 weeks were transferred to the control diet for an additional 36 weeks. The Nv again decreased over this time from 1.08 to 0.133 $mm^3$. The corresponding Vq showed an increase, with the individual mean volume for a focus increasing from 0.025 to 0.136 $mm^3$. These data for foci development show a decrease in number after an apparent maximum is reached, along with an increase in size of remaining foci. In all of the groups analyzed, areas of hepatocytomegally disappeared as the time on the control diet increased. These areas were completely reversible.

Quantitative Evaluation of HA and HPC

In general, the total number of neoplasms (HA and HPC) and the percentage of neoplasms that were HPC increased as the time interval from initiation of safrole feeding to killing and autopsy increased. This trend was seen in the continuous exposure groups (16-52 weeks) and in all groups removed from the safrole diet and placed on the

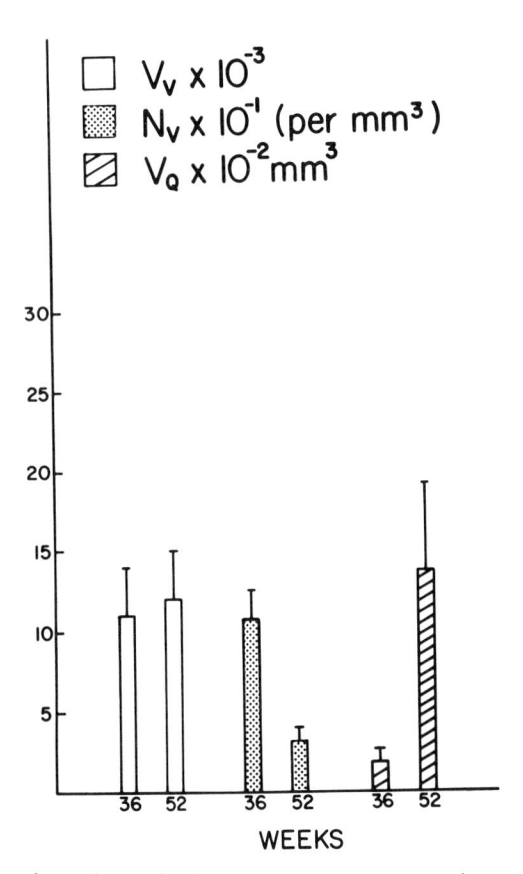

FIGURE 6. Plot of the morphometric data (mean ± SEM) for groups 4 and 6, fed safrole continuously for 36 and 52 weeks, respectively.

Legend: $V_V \times 10^{-3}$; $N_V \times 10^{-1}$ (per mm$^3$); $V_Q \times 10^{-2}$ mm$^3$

control diet for an additional 36 or 52 weeks. Table 4 summarizes the distribution of total neoplasms within the groups, according to the duration of safrole exposure. No mice developed neoplasms after 16 weeks of continuous safrole feeding. However, when mice were allowed to live an additional 36 weeks on the control diet, 15% developed grossly visible neoplasms. This increased to 50% in mice allowed to live for 52 wk on the control diet subsequent to a 16 wk safrole exposure. The number of neoplasms per mouse remained the same at each time point.

After 36 weeks of continuous safrole feeding, 45.8% developed neoplasms. This increased to 100% when the 36 weeks of safrole feeding was followed by an additional 36 weeks of control diet. In these groups, the number of neoplasms per mouse increased from 1.8 to 5.7. A similar increase in the percent incidence and neoplasms per mouse was seen in the 52-week safrole group and in the 52-week safrole plus 36-week control group. The pattern demonstrated by the incidence data was also apparent in the size distribution of the

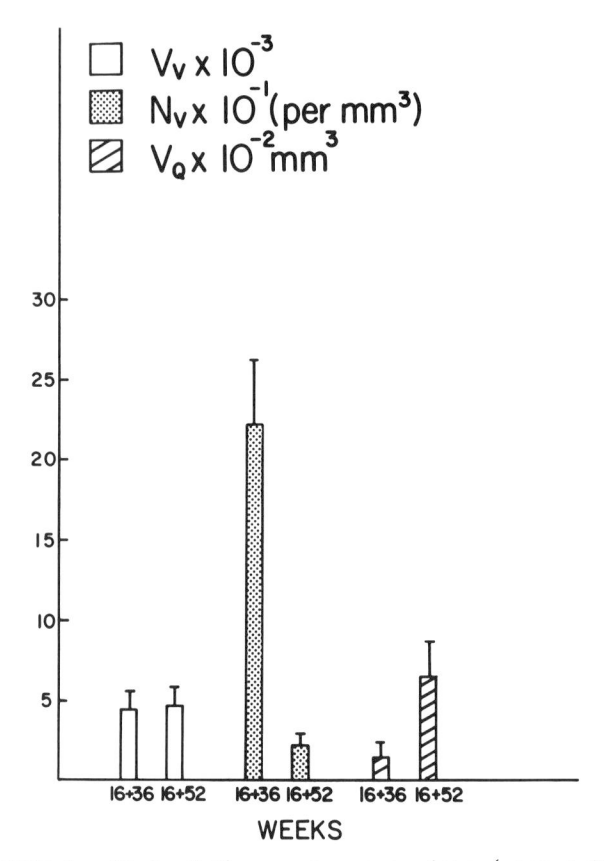

FIGURE 7. Plot of the morphometric data (mean ± SEM) for groups 2 and 3, fed safrole for 16 weeks, then fed control diet for 36 and 52 weeks, respectively.

neoplasms. In general, the mean diameter, the range of diameters, and percentage of neoplasms greater than 10 mm in diameter increased with time, even in the absence of safrole treatment. These data are presented in Table 5. In the groups fed safrole for 16 weeks followed by the control diet for 36 and 52 weeks, the mean diameter of the neoplasms measured increased from 1.9 to 3.9 mm. The percentage of neoplasms with a diameter greater than 10 mm increased from 0 to 6%. A similar trend was noted in mice on the 36- and 52-week safrole regimens. The mean diameter for neoplasms in mice fed safrole for 52 weeks followed by the control diet for 36 weeks was 8.6 mm, and 44% were greater than 10 mm in diameter.

The morphological classification of all neoplasms and their distribution among the groups is presented in Table 6. The total number of neoplasms increased with time, even in the absence of safrole treatment. In general, the percentage of HPC also increased with time, except in the 16-week-safrole-fed groups. At 36 weeks,

SAFROLE-INDUCED MOUSE LIVER LESIONS 171

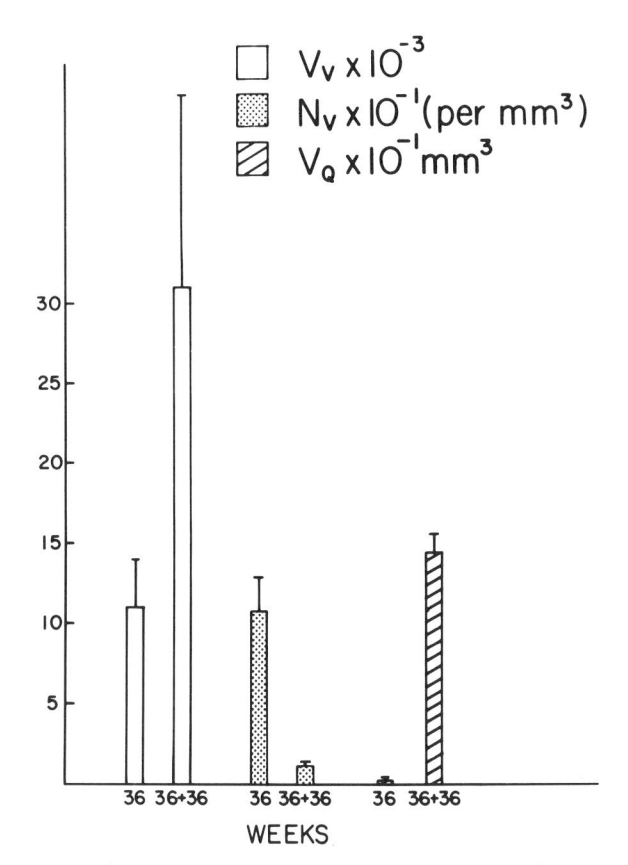

FIGURE 8. Plot of the morphometric data (mean $\pm$ SEM) for group 4, fed safrole for 36 weeks, and group 5, fed safrole for 36 weeks followed by 36 weeks of the control diet.

the HPC comprised 4.8% of all neoplasms. After an additional 36 weeks on the control diet, this increased to 22%. Only 1 of the control mice (1/61) developed neoplasms at 90 weeks. This mouse had 7 HA and 1 HPC.

DISCUSSION

The overall goal of research in our laboratory has been to understand the biological nature, behavior, and, thereby, the significance of chemically-induced hepatocellular neoplasia in the mouse. Our previous work, utilizing continuous dietary administration of safrole (4000 ppm w/w) as a model HPC-inducing regimen, identified a number of hepatocellular alterations that developed in a sequential manner prior to HPC (3). We subsequently characterized these lesions by light and electron microscopy, enzyme histochemistry, transplantation, and biochemical techniques (1-4). The malignant

TABLE 4. Distribution of Neoplasms

| Weeks | Neoplasms/ mouse | SD | Variance | Mice with neoplasms | Number of mice |
|---|---|---|---|---|---|
| 16 | - | - | - | - | 25 |
| 16 S+36 cont. | 2.8 | ± 2.89 | 5.6 | 15 | 20 |
| 16 S+52 cont. | 2.4 | ± 1.27 | 1.4 | 50 | 14 |
| 36 S | 1.8 | ± 1.32 | 1.6 | 45.8 | 24 |
| 36 S+36 cont. | 5.7 | ± 3.05 | 8.7 | 100 | 16 |
| 52 S | 3.1 | ± 2.52 | 5.6 | 77 | 22 |
| 52 S+36 cont. | 4.9 | ± 2.14 | 4.2 | 93 | 14 |

Controls: 24-52 weeks (0/40), 90 weeks (1/21), 8 neoplasms

S = Safrole feeding; cont. = control diet feeding.

TABLE 5. Size of Neoplasms

| Weeks of protocol | Mean diameter (mm) | SD | Variance | Range (mm) |
|---|---|---|---|---|
| 16 S | - | - | - | - |
| 16 S+36 cont. | 1.9 | ± 1.46 | 1.8 | 1-5 |
| 16 S+52 cont. | 3.9 | ± 3.87 | 14.07 | 1-17 (6% > 10 mm) |
| 36 S | 2.2 | ± 2.08 | 4.10 | 0.5-7 |
| 36 S+36 cont. | 5.7 | ± 4.19 | 17.4 | 1-18 (18% > 10 mm) |
| 52 S | 4.2 | ± 3.03 | 9.0 | 1-10 (6% > 10 mm) |
| 52 S+36 cont. | 8.6 | ± 4.77 | 22.4 | 1-22 (44% > 10 mm) |

S = Safrole feeding; cont. = control diet feeding.

TABLE 6. Classification of Neoplasms

| Weeks of protocol | Total number of neoplasms | HA (%) | Number HA | HPC (%) | Number HPC |
|---|---|---|---|---|---|
| 16 S | 0 | 0 | 0 | 0 | 0 |
| 16 S+36 cont. | 3 | 33.3 | 1 | 67.7 | 2 |
| 16 S+52 cont. | 15 | 86.7 | 13 | 13.3 | 2 |
| 36 S | 21 | 95.2 | 20 | 4.8 | 1 |
| 36 S+36 cont. | 81 | 78 | 63 | 22 | 18 |
| 52 S | 52 | 75 | 39 | 25 | 13 |
| 52 S+36 cont. | 64 | 61 | 39 | 39 | 25 |
| All controls | 8 | – | 7 | – | 1 |

S = Safrole feeding; cont. = control diet feeding.

nature of the HPC was demonstrated by their ability to metastasize and to transplant into syngeneic hosts. The behavior and significance of foci and HA, however, remained in question. Most of these lesions were similar, in a variety of ways, to chemically induced hepatic lesions in rats studied by a large number of investigators (15-20). Recent reports have also described characteristics of mouse liver foci and HA induced by different classes of chemical carcinogens, and the characteristics are similar to those noted in our studies (21-24). Our studies demonstrated the usefulness of potential enzymatic markers, including G-6-pase and GGT, in the identification of phenotypic alterations in hepatocytes of carcinogen-treated mice. However, there is still little data relating these phenotypic changes to functional importance in the carcinogenic process in mouse liver (21).

The quantitative experiments presented in this report were designed to gain insight into the importance of foci and HA in HPC development in mice. A number of investigators have established the usefulness of quantitative morphology (stereology) in delineating the dynamics of hepatocellular carcinogenesis in rats (25-29). In the present work, we used GGT-positive reaction as a single marker for phenotypic alteration to perform the necessary measurements for the stereologic estimations. This marker uniformly labeled over 85% of all available foci, as determined by our initial investigations. This 85% was assumed to be characteristic of the total number of foci present in the livers, and the behavior of these was assumed to be representative of that total population. Our data indicate that most of the foci that developed during the different safrole-feeding schedules were probably reversible, i.e., reverted to a normal phenotypic profile after removal of the carcinogenic stimulus. This was evidenced by a decrease in Nv. The foci that remained, however, increased in size, as evidenced by a larger Vq, such that the relative volume (Vv) of liver composed of GGT-positive foci did not change after safrole was removed from the diet. The most feasible explanation is that the persisting induced foci grew autonomously in the absence of safrole. Based on the Nv values, only 7-10% of the foci marked by GGT were persistent. The distribution and wide separation of foci within the liver lobes makes it unlikely that the larger foci seen at the various endpoints of our protocol were due to coalescence of the many smaller foci. Vesselinovitch and Mihailovich (30) have demonstrated mathematically that coalescence is an unlikely explanation for similar results in their study. In the present study, we were unable to distinguish, in the earlier samples, which foci were capable of persistence and growth and which foci were destined to reverse. The incorporation of multiple enzymatic and functional markers into the quantitative analysis may provide the means to associate phenotype with behavior. Pugh and Goldfarb (29) showed that, in rat liver, foci with the widest diversity of phenotypic expression, as defined by alterations in 3 marker enzymes, also contained the highest level of DNA labeling with tritiated thymidine. At present, with our model, the most we can say is that a majority of foci that develop in mouse liver after continuous safrole feeding revert to a normal phenotype (as defined by light microscopy) after cessation of exposure. Some of the foci, however, persist and have developed autonomous growth potential. These foci increase in size and must be considered as putative precursors of HA and HPC.

The HA and HPC induced by safrole appear to persist, with the HA capable of growth and possible progression even in the absence of the safrole stimulus. In all of our experimental group, the HA and HPC incidence, the number of neoplasms per mouse, and the mean diameter of neoplasms increased with time, even in the absence of safrole treatment. The data in Table 6 showed a decrease in the HA incidence at later sampling points corresponding to an increase in the HPC incidence. This may indicate a slow rate of conversion of some HA to HPC. The development in and subsequent overgrowth of mouse HA by HPC cells was noted in our previous work and by others (3,31,32). From the incidence rates, it can be estimated that this regression occurs in a relatively small number of HA. Most of the HA, however, seem capable of continued growth and development, which is a strong indication that these lesions are neoplastic. A similar conclusion was reached by Hirota and Williams (8) when they investigated carcinogen-induced neoplastic nodules in rat liver. However, they considered progression to HPC a rare phenomenon. They also considered foci to be precursors to the nodules. Our data shows a decrease in the Nv of foci, and the increase in Vq corresponded to the increased development of HA (Tables 3 and 4). This may be consistent with a precursor relationship of foci to HA. Ito et al. (11) reported the existence of reversible and irreversible nodules in 1,2,3,4,5,6-hexachlorocyclohexane-treated mice. The original morphology, however, was identical for both types of lesions. Although our data does not appear to support reversibility of HA, it cannot be ruled out. Another possible explanation for the persistence, growth, and progression of foci and HA after cessation of safrole feeding is the presence of an unidentified promoter in our control diet. This possibility seems minimal, however, since only 1 of 91 control mice developed liver neoplasms. A promoter would be expected to increase the background spontaneous tumor incidence.

In summary, it is apparent that only 16 weeks of safrole treatment is needed to produce initiated hepatocytes that are far enough along in the neoplastic process to progress to HPC without further carcinogenic stimulation. Most hepatocellular foci seem to be reversible, yet perhaps as many as 10% persisted and grew. The foci must be considered as putative precursors of HA and probably HPC in the mouse. Similarly, the HA appear to possess autonomous growth capability with some percentage progressing to HPC. More work is needed to assess the functional significance of different enzymatic markers and to trace individual cell populations from initiation to HPC development, to conclusively establish the geneology of the HPC cells.

REFERENCES

1. Lipsky MM, Hinton DE, Klaunig JE, Goldblatt PJ, Trump BF: Iron-negative foci and nodules in safrole-treated mouse liver made siderotic by iron-dextran injection. Pathol Res Pract 164:178-185, 1979.

2. Lipsky MM, Hinton DE, Klaunig JE, Trump BF: Biology of hepatocellular neoplasia in the mouse. I. Histogenesis of safrole-induced hepatocellular carcinoma. J Natl Cancer Inst 67:365-376, 1981.

3. Lipsky MM, Hinton DE, Klaunig JE, Goldblatt PJ, Trump BF: Biology of hepatocellular neoplasia in the mouse. II. Sequential enzyme histochemical analysis of Balb/c mouse liver during safrole-induced carcinogenesis. J Natl Cancer Inst 67:377-392, 1981.

4. Lipsky MM, Hinton DE, Klaunig JE, Trump BF: Biology of hepatocellular neoplasia in the mouse. III. Electron microscopy of safrole-induced hepatocellular adenomas and hepatocellular carcinomas. J Natl Cancer Inst 67:393-405, 1981.

5. Homburger F, Boger E: The carcinogenicity of essential oils, flavors and species. A review. Cancer Res 28:2372-2374, 1968.

6. Homburger F, Kelley T, Baker TR, Russfield AB: Sex effects on hepatic pathology from deficient diet and safrole in rats. Arch Pathol 73:118-125, 1962.

7. Long EL, Nelson AA, Fitzhugh OG, Hansen W: Liver tumors produced in rats by feeding safrole. Arch Pathol 75:595-603,

1963.

8. Borchert P, Miller JA, Miller EC, Shires TK: 1'-Hydroxysafrole, a proximate carcinogenic metabolite of safrole in rat and mouse. Cancer Res 33:590-600, 1973.

9. Ito N, Hananouchi M, Sugihara S, Shirai S, Shirai T, Tsuda H, Fukushia S, Nagasaki H: Reversibility and irreversibility of liver tumors in mice induced by the isomer of 1,2,3,4,5,6-hexachlorocyclohexane. Cancer Res 36:2227-2234, 1976.

10. Hirota N, Williams GM: Persistence and growth of rat liver neoplastic nodules following cessation of carcinogen exposure. J Natl Cancer Inst 63:1257-1265, 1979.

11. Ruttenberg AM, Kim H, Fishbein JW, Hanker LS, Wasserburg HI, Seligman AM: Histochemical and ultrastructural demonstration of gamma-glutamyl transpeptidase activity. J Histochem Cytochem 17:517-526, 1969.

12. Wachstein M, Meisel E: Histochemistry of hepatic phosphatases at a physiologic pH. Am J Clin Pathol 27:13-23, 1957.

13. Negi DS, Stephen RJ: An improved method for the histochemical localization of glucose-6-phosphate dehydrogenase in animal and plant tissue. J Histochem Cytochem 25:149-154, 1977.

14. Underwood EE: Quantitative Stereology. Reading, Pa.: Addison Wesley, 1970.

15. Farber E: Hyperplastic areas, hyperplastic nodules and hyperbasophilic areas as putative precursor lesions. Cancer Res 36:2532-2533, 1976.

16. Reuber MD: Development of preneoplastic and neoplastic lesions of the liver in male rats given 0.025 percent $\underline{N}$-2-fluorenyldiacetamide. J Natl Cancer Inst 34:697-724, 1965.

17. Epstein S, Ito N, Merkow L, Farber E: The cellular analysis of liver carcinogenesis. I. The induction of large hyperplastic nodules in the liver with 2-fluorenylacetamide or ethionine and some aspects of their morphology and glycogen metabolism. Cancer Res 27:1712-1721, 1967.

18. Newberne PM, Wogan GM: Sequential morphologic changes in aflatoxin $B_1$ carcinogenesis in the rat. Cancer Res 28:770-781, 1968.

19. Williams GM: Functional markers and growth behavior of preneoplastic hepatocytes. Cancer Res 36:2540-2543, 1976.

20. Goldfarb S: A morphological and histochemical study of carcinogenesis of the liver in rats fed 3'-methyl-4-dimethylaminoazobenzene. Cancer Res 33:1119-1128, 1973.

21. Moore MM, Drinkwater NR, Miller EC, Miller JA, Pitot HC: Quantitative analysis of the time-dependent development of glucose-6-phosphatase-deficient foci in the livers of mice treated neonatally with diethylnitrosamine. Cancer Res 41:1585-1593, 1981.

22. Jalanko H, Ruoslahti E: Differential expression of α-fetoprotein and γ-glutamyl transpeptidase in chemical and spontaneous hepatocarcinogenesis. Cancer Res 39:3495-3501, 1979.

23. Frith CH, Baetcke KP, Nelson CJ, Schieferstein G: Sequential morphogenesis of liver tumors in mice given benzidine dihydrochloride. Eur J Cancer 16:1205-1216, 1980.

24. Reubner BH, Gershwin ME, French SW, Meierhenry E, Dunn P, Hsieh LS: Mouse hepatic neoplasia: Differences among strains and carcinogens. Chap. 9, this volume.

25. Scherer E, Emmelot P: Kinetics of induction and growth of precancerous liver cell foci and liver tumor formation by diethylnitrosamine in the rat. Eur J Cancer 11:689-701, 1975.

26. Emmelot P, Scherer E: The first relevant cell stage in rat liver carcinogenesis: A quantitative approach. Biochim Biophys Acta 605:247-304, 1980.

27. Pitot HC, Barsness L, Goldsworthy T, Kitagawa J: Biochemical characterization of stages of hepatocarcinogenesis after a single dose of diethylnitrosamine. Nature (Lond) 271:456-458, 1978.

28. Pitot HC, Goldsworthy T, Campbell HA, Poland A: Quantitative evaluation of the promotion by 2,3,7,8-tetrachlorodibenzo-p-dioxin of hepatcarcinogenesis from diethylnitrosamine. Cancer Res 40:3616-3620, 1980.

29. Pugh TD, Goldfarb S: Quantitative histochemical and autoradiographic studies of hepatocarcinogenesis in rats fed 2-acetylaminofluorene followed by phenobarbital. Cancer Res 38:4450-4457, 1978.

30. Vesselinovitch SD, Mihailovich N: Kinetics of induction and growth of basophilic foci and development of hepatocellular carcinoma by diethylnitrosamine in the infant mouse. Chap. 5, this volume.

31. Frith CH, Dooley K: Brief communication: Hepatic cytologic and neoplastic changes in mice given benzidine dihydrochloride. J Natl Cancer Inst 56:679-682, 1976.

32. Ward JM, Vlahakis G: Evaluation of hepatocellular neoplasms in mice. J Natl Cancer Inst 61:807-811, 1978.

# Index

2-Acetylaminofluorene-induced neoplasms, 17, 85–93, 145, 153
  biologic characteristics, 87
  initiation-promotion studies, 108
  morphology, 86, 90, 91
Adenoma (*see* Hepatocellular adenoma)
Adenosine triphosphatase reaction, in DEN- and dieldrin-induced neoplasms, 132, 133
Aldrin-induced neoplasms, 30
Alkaline phosphatase reaction, in DEN- and dieldrin-induced neoplasms, 132, 133
Alkylating agents:
  carcinogenesis, 95–102
  cell replication and, 96–97
  DNA damage and repair and, 95–96
$O^6$-Alkylguanine, and carcinogenesis, 95–96
Alpha-fetoprotein, and hepatotumorigenesis, 147, 150
2-Aminoanthraquinone-induced tumors, 32

3-Amino-4-ethoxyacetamide-induced tumors, 32
3-Amino-9-ethylcarbazole hydrochloride-induced tumors, 32
1-Amino-2-methylanthraquinone-induced tumors, 32
4-Amino-2-nitrophenol-induced tumors, 32
2-Amino-5-nitrothiazole-induced tumors, 32
Aniline hydrochloride-induced tumors, 32
*o*-Anisidine hydrochloride-induced tumors, 32
Azobenzene-induced tumors, 32

Basophilic foci, DEN-induced, 61, 65–66, 80
  dose vs time to 50% incidence, 69–73, 74
  dose vs transformation possibility, 68, 69, 70
  size, number of cells, and percent of lesions, 78

# INDEX

Basophilic hepatocytes, 3
  in dieldrin- and DEN-induced neoplasms, 128
B6C3F1 mice:
  DEN-induced neoplasms, 120, 122, 123, 128
    basophilic vs eosinophilic, 126
    growth curves and liver weights, 119
  dieldrin-induced neoplasms, 120, 122, 123, 125, 128
    basophilic vs eosinophilic, 126
    growth curves and liver weights, 119
  NCI chemical carcinogen bioassays, 27-37
α-Benzene hexachloride-induced neoplasms, 17
Benzidine dihydrochloride-induced neoplasms, 85-93
  biologic characteristics, 87
  morphology, 86, 90, 91
Benzidine-induced neoplasms, 17
Bioassays, NCI, 27-37

Captan-induced neoplasms, 30
Carbon tetrachloride, as promoting agent, 111
Carcinogenesis:
  alkylating agents, 95-102
    cell replication in initiation and promotion, 96-97
    cell specificity, 97-102
    DNA alkylation and repair, 95-96
    initiation index, 99
  initiation and promotion studies, 107-113
    cell replication in, 96-97
  (See also Chemical carcinogens)
Castration, tumor incidence and, 54, 55, 56
C57BL/6 mice:
  DEN-induced neoplasms, 120, 128
    basophilic vs eosinophilic, 126
    growth curves and liver weights, 117
  dieldrin-induced neoplasms, 120, 128, 129
    basophilic vs eosinophilic, 126
    growth curves and liver weights, 117
Cell replication, in initiation and promotion of carcinogenesis, 96-97
C3H/He mice:
  DEN-induced neoplasms, 120, 121, 123, 124, 128
    basophilic vs eosinophilic, 126
    growth curves and liver weights, 118

  dieldrin-induced neoplasms, 120, 121, 123, 128
    basophilic vs eosinophilic, 126
    growth curves and liver weights, 118
Chemical carcinogens, 1, 11, 13, 15, 17, 30-34
  (See also Neoplastic lesions, chemically induced; specific carcinogens)
  2-acetylaminofluorene, 85-93
  alkylating agents, 95-102
  benzidine dihydrochloride, 85-93
  dieldrin, 115-141
  diethylnitrosamine, 61-80, 115-141
  epigenetic, 11, 13, 17
  genotoxic, 13, 17
  human risk assessment, 39-44
  NCI bioassays, 27-37
  promotor, 108
  safrole, 161-175
Chloramben-induced neoplasms, 30
Chlordane-induced neoplasms, 17, 30, 145, 153
Chlordecone-induced neoplasms, 30
Chlorobenzilate-induced neoplasms, 30
Chloroform-induced neoplasms, 17, 30
3-(Chloromethyl) pyridine hydrochloride-induced neoplasms, 32
4-Chloro-m-phenylenediamine-induced neoplasms, 17, 32
4-Chloro-o-phenylenediamine-induced neoplasms, 32
Chlorothalonil-induced neoplasms, 30
4-Chloro-o-toluidine hydrochloride-induced neoplasms, 32
5-Chloro-o-toluidine-induced neoplasms, 32
Cholangiosarcoma, 20
C.I. vat yellow 4-induced neoplasms, 32
Cinnamyl anthranilate-induced neoplasms, 32
Classification, 1-2, 147-150, 154-156
  chemically induced tumors, 11
  morphology, 147-150, 154-156
m-Cresidine-induced neoplasms, 32
p-Cresidine-induced neoplasms, 32
Cupferron-induced neoplasms, 32

Dapsone-induced neoplasms, 32
p,p'-DDE-induced neoplasms, 32
DEN, (see Diethylnitrosamine)
2,4-Diaminoanisole-induced neoplasms, 32
2,4-Diaminotoluene-induced neoplasms, 32
Diaminozide-induced neoplasms, 32

# INDEX

Dibromochloropropane-induced neoplasms, 30
1,2-Dibromoethane-induced neoplasms, 30
1,2-Dichloroethane-induced neoplasms, 30
Dicofol-induced neoplasms, 30
Dieldrin-induced neoplasms, 17, 115-141
    electron microscopy, 134-137, 138, 139, 140
    enzyme histochemistry, 128, 131, 132, 133
    growth curves and liver weights, 117-119
    light microscopy, 120-128
    strain differences, 115-141
Diet:
    hepatic neoplasms and, 153-154
    tumor incidence and, 48-51, 55-58
Di(2-ethylhexyl)phthalate-induced neoplasms, 17
Diethylnitrosamine (DEN)-induced neoplasms, 17, 61-80, 145
    electron microscopy, 134-137, 140
    enzyme histochemistry, 128, 131, 132, 133
    growth curves and liver weights, 117-119
    initiation-promotion studies, 108
    light microscopy, 120
    strain differences, 115-141
Diethylstilbestrol (DES), tumor incidence and, 54, 55
$N,N'$-Diethylthiourea-induced neoplasms, 32
3,3'-Dimethoxybenzidine-4,4'-diisocyanate-induced neoplasms, 33
Dimethylnitrosamine (DMN)-induced neoplasms, 96, 97-102
    phenobarbital enhancement, 112
2,4-Dinitrotoluene-induced neoplasms, 33
1,4-Dioxane-induced neoplasms, 30
Direct black 38-induced neoplasms, 33
Direct blue 6-induced neoplasms, 33
Direct brown 95-induced neoplasms, 33
DNA alkylation, 95-96
DNA repair, 95-96
DNA synthesis, DMN exposure and, 99-100

Endocrine factors, in tumor incidence, 53-55, 56-58
Enzyme histochemistry:
    in dieldrin- and DEN-induced tumors, 128, 131, 132, 133
    in safrole-induced neoplasms, 163, 167

Eosinophilic hepatocytes, 4
    in dieldrin- and DEN-induced neoplasms, 137-139
Epigenetic carcinogens, 11, 13, 17
Estradiol, tumor incidence and, 54-55
Estrone, tumor incidence and, 54-55
Ethyl tellurac-induced neoplasms, 33

Fats, dietary, tumor incidence and, 50
Fischer 344 (F-344) rats, NCI chemical carcinogen bioassays, 27, 29, 32-34, 35, 36, 37
$N$-2-Fluorenylacetamide (2-AAF) (*see* 2-Acetylaminofluorene)
Focal hyperplasia, 3
    terminology, 12
Foci, 3-5
    in safrole-induced neoplasms, 163, 166-168
    trabecular, 12, 14

Gamma-glutamyl transpeptidase, in safrole-induced neoplasms, 163, 164, 166, 167
Genetic factors, tumor incidence and, 51-53, 56-58
Genotoxic carcinogens, 13, 17
Glucocorticoids, hepatic effects, 53
Glucose-6-phosphatase reaction:
    in dieldrin- and DEN-induced neoplasms, 132, 133
    in safrole-induced neoplasms, 163, 165, 167
Glucose-6-phosphate dehydrogenase, in safrole-induced neoplasms, 163, 164, 167
Glycogen, in dieldrin- and DEN-induced neoplasms, 132, 133
Growth hormone, tumor incidence and, 55

Hemangiosarcoma, 20
Hepatoblastoma, 15, 20
Hepatocellular adenoma, 6
    2-acetylaminofluorene-induced, 87
    basophilic cells, 88
    benzidine-induced, 87
    carcinoma arising in, 8-9, 12
    DEN-induced, 124
        biologic behavior, 64
        dose vs time to 50% incidence, 74

# INDEX

Hepatocellular adenoma, DEN-induced (*Cont.*):
  dose vs transformation possibility, 68, 69
  enzyme histochemistry, 128, 132, 133
  size, number of cells, and percent of lesions, 78
  transplantability, 62, 63
  diagnostic criteria, 62
  dieldrin-induced, 125
  enzyme histochemistry, 128, 132, 133
  evidence for neoplasia, 13
  histopathology, 6-8, 9, 10, 11
  safrole-induced, 161-175
  quantitative evaluation, 168-171, 173-175
  trabecular foci in, 12, 14
Hepatocellular carcinoma:
  2-acetylaminofluorene-induced, 85-93
  alkylating agents and, 95-102
  basophilic cells, 89
  benzidine-induced, 85-93
  acidophilic cells, 89
  DEN-induced, 61-80, 123, 127, 128
  biologic behavior, 64
  dose vs growth, mitotic activity, and development, 73-75, 76, 77, 78, 79
  dose vs time to 50% incidence, 69-73, 74
  dose vs transformation possibility, 68, 69
  enzyme histochemistry, 128, 131, 132, 133
  size, number of cells, and percent of lesions, 78
  transplantability, 62, 63
  diagnostic criteria, 62
  dieldrin-induced, 123, 128
  enzyme histochemistry, 128, 132, 133
  dimethylnitrosamine-induced, 96, 97-102
  DNA alkylation and cell replication in, 95-102
  in hepatocellular adenoma, 8-9, 12
  histopathology, 9
  metastasis, 9, 14, 17
  phenobarbital exposure and, 152, 155
  safrole-induced, 161-175
  quantitative evaluation, 168-171, 173-175

  trabecular, 16
Hepatocellular neoplasms (*see* Neoplastic lesions; *specific neoplasms*)
Hepatocytes:
  basophilic, 3, 136
  in dieldrin-induced neoplasms, 121-122, 124, 125, 128, 134-135
  in diethylnitrosamine-induced neoplasms, 136
  eosinophilic, 4
  in foci, 3-5
Heptachlor-induced neoplasms, 30
Hexachloroethane-induced neoplasms, 30
Histiocytic sarcoma, 20
Histogenesis, 3-5
Hormones, tumor incidence and, 53-55, 56-58
Hydrazobenzene-induced neoplasms, 33
Hyperplastic nodules:
  DEN-induced, 61
  biologic behavior, 64
  dose vs time to 50% incidence, 74
  dose vs transformation possibility, 68, 69
  size, number of cells, and percent of lesions, 78
  transplantability, 62, 63
  diagnostic criteria, 62

Initiation (*see* Carcinogenesis, initiation and promotion)
Initiation index, 99

Liver (*see* Hepato entries)

Mallory body(ies), in dieldrin-induced neoplasms, 128, 129, 130
Metastasis, 9, 14, 17
4,4'-Methylene-bis(*N,N*-dimethyl)-benzenamine-induced neoplasms, 33
$O^6$-Methylguanine, carcinogenesis, 96
Methylnitrosourea (MNU)-induced neoplasms, 96
Michler's ketone-induced neoplasms, 33

1,5-Naphthalenediamine-induced neoplasms, 33
National Cancer Institute (NCI) bioassays, 27-37

# INDEX

National Cancer Institute (NCI) bioassays, (*Cont.*):
  experimental design, 27-28
  materials (animals), 27
Neoplastic lesions:
  basophilic, chemicals inducing, 17
  chemically induced, 11, 13, 15, 145-156
    2-acetylaminofluorene, 85-93
    alpha-fetoprotein in, 147, 150
    benzidine hydrochloride, 85-93
    dieldrin, 115-141
    diethylnitrosamine (DEN), 61-80, 115-141
    hormonal influences, 56
    implications for predicting carcinogenesis in humans, 39-44
    morphology, 147-150
    NCI bioassays, 27-37
    safrole-induced, 161-175
    transplantation studies, 147, 150-151
    virus studies, 151
  classification, 1-2, 147-150, 154-156
  diet and, 48-51, 55-58, 153-154
  eosinophilic, chemicals inducing, 17
  etiology in various species, 5
  genetic factors, 51-53, 56-58
  histogenesis, 3-5
  histopathogenesis (potential), 3
  hormonal factors, 53-55, 56-58
  initiation-promotion (*see* Carcinogenesis)
  morphologic classification, 147-150, 154-156
  nonhepatocellular, 15, 18, 20
  pathology in various species, 5
  phenobarbital exposure and, 152-153, 155
  spontaneous (naturally occurring), 3-5, 145-156
    alpha-fetoprotein in, 147, 150
    biologic characteristics, 87
    morphologic characteristics, 86, 88, 90, 147-150
    transplantation studies, 147, 150-151
    virus studies, 151
  terminology, 12
  trabeculation, 149
  (*See also specific neoplasms*)
Nithizide-induced neoplasms, 33
5-Nitroacenaphthene-induced neoplasms, 33

3-Nitro-*p*-acetophenetide-induced neoplasms, 33
5-Nitro-*o*-anisidine-induced neoplasms, 33
6-Nitrobenzimidazole-induced neoplasms, 33
Nitrofen-induced neoplasms, 17, 30, 33
Nitrolotriacetic acid-induced neoplasms, 33
2-Nitro-*p*-phenylenediamine-induced neoplasms, 17, 33
3-Nitropropionic acid-induced neoplasms, 33
*N*-Nitrosodiphenylamine-induced neoplasms, 33
*p*-Nitrosodiphenylamine-induced neoplasms, 33
5-Nitro-*o*-toluidine-induced neoplasms, 33

Osborne Mendel (OM) rats, NCI chemical carcinogen bioassays in, 27, 29-31, 35, 36, 37

Phenazopyridine hydrochloride-induced neoplasms, 33
Phenobarbital, as promoting agent, 17, 108, 109-111, 112, 152-153, 155
Phenobarbitone, as promoting agent, 111
Piperonyl sulfoxide-induced neoplasms, 33
Pivalolactone-induced neoplasms, 33
Pregnancy, tumor incidence and, 54, 55
Preneoplastic (potential) lesions:
  classification, 2
  histogenesis, 3-5
Progesterone, hepatic effects, 53
Promotion (*see* Carcinogenesis, initiation and promotion)
Proteins, dietary, tumor incidence and, 49-50

*p*-Quinone dioxime-induced neoplasms, 34

Reserpine-induced neoplasms, 34

Safrole-induced neoplasms, 161-175
  enzyme histochemistry, 163, 167
  morphologic alterations, 163-166
  morphometric analysis of foci, 166-168

Safrole-induced neoplasms (*Cont.*):
   quantitative evaluation of HA and HPC, 168–171, 172, 173
Selenium sulfide-induced neoplasms, 34
Succinic dehydrogenase reaction:
   in DEN-induced neoplasms, 131, 132, 133
   in dieldrin-induced neoplasms, 132, 133
Sulfallate-induced neoplasms, 30

Testosterone, tumor incidence and, 54, 55
1,1,2,2-Tetrachloroethane-induced neoplasms, 30
Tetrachloroethylene-induced neoplasms, 30
Tetrachlorvinphos-induced neoplasms, 17, 19, 30
4,4′-Thiodianiline-induced neoplasms, 34
Thyroxine, tumor incidence and, 55
*o*-Toluidine hydrochloride-induced neoplasms, 34
Toxophene-induced neoplasms, 17, 18, 30
Trabeculation, 90, 92, 149
Transplantation studies:
   DEN-induced neoplasms, 62–63, 65
   hepatotumorigenesis, 147, 150–151
1,1,2-Trichloroethane-induced neoplasms, 30
Trichloroethylene-induced neoplasms, 31
2,4,6-Trichlorophenol-induced neoplasms, 34
Trifuralin-induced neoplasms, 31
2,4,5-Trimethylaniline-induced neoplasms, 34
Trimethyl phosphate-induced neoplasms, 34
Trimethylthiourea-induced neoplasms, 34
Tris(2,3-dibromopropyl)phosphate-induced neoplasms, 34

Virus studies, in hepatic neoplasia, 151